Bubbles and Bonanzas

Bubbles and Bonanzas

British Investors and Investments in Mexico, 1821–1860

Michael P. Costeloe

LEXINGTON BOOKS
A division of
ROWMAN & LITTLEFIELD PUBLISHERS, INC.
Lanham • Boulder • New York • Toronto • Plymouth, UK

Published by Lexington Books
A wholly owned subsidiary of The Rowman & Littlefield Publishing Group, Inc.
4501 Forbes Boulevard, Suite 200, Lanham, Maryland 20706
www.lexingonbooks.com

Estover Road, Plymouth PL6 7PY, United Kingdom

British Library Cataloguing in Publication Information Available

Library of Congress Cataloging-in-Publication Data
Costeloe, Michael P.
Bubbles and bonanzas : British investors and investments in Mexico, 1821–1860 / Michael P. Costeloe.
p. cm.
Includes bibliographical references and index.
ISBN 978-0-7391-5119-8 (cloth : alk. paper) — ISBN 978-0-7391-5121-1 (electronic)
1. Investments, British—Mexico—History—19th century. 2. Silver mines and mining—Mexico—History—19th century. 3. Mineral industries—Mexico—History—19th century. 4. Mexico—History—1821–1861. I. Title.
HG5162.C67 2011
332.67'34107209034—dc22 2011003742

™ The paper used in this publication meets the minimum requirements of American National Standard for Information Sciences—Permanence of Paper for Printed Library Materials, ANSI/NISO Z39.48-1992.

Printed in the United States of America

Contents

Acknowledgments

Grateful acknowledgment is made to the many scholars and to the staffs of national and local archives, and of public and university libraries in the United Kingdom, the United States, Australia, and Mexico, without whose assistance and cooperation the research for this book would not have been possible. I am particularly indebted to Michael Hironymous, Anne Staples, Flor Salazar, Linda Arnold, Margaret Ashby, Melisssa Boyd, Josefina Vázquez, Roger Kerrison, Brian Mills, Nicholas Webb, William Reese, Neil Boness, Sara Hilder, George Gorzynski, Rosalina Ríos Zuñiga, Belem Oviedo, Michael Treays, Marigel Orozco Guzmán, María Lucila López de la Vega, William Fowler, Edgar Omar Gutiérrez, and Alma Parra. They all provided, freely and generously, copies of documents as well as information essential for my research. As always, my wife Eleanor supplied indispensable support with the research on many occasions, working in archives, copying texts, checking sources, and in every way giving the help that I needed. To all those named, and to those I have inadvertently omitted, I am much indebted.

Introduction

History is about evidence: no evidence, no history.[1]

It is a little-known fact of British history that in the two years between 1824 and 1825, British people invested approximately ten million pounds in Mexico. If we accept the calculation of recent research that the purchasing power of £1 in 1824 is now in the year 2010 about seventy times more, then the sum invested in our terms amounted to £700 million pounds.[2] Adding in other costs, fees, and miscellaneous extras with several millions more invested at the same time elsewhere in Latin America, the total investment amounted to more than one billion pounds.

This enormous sum of money did not come from the British government, or the Bank of England, or other banks and financial institutions. Nor for the most part was it taken from the deep pockets of rich merchants and financiers, landowners or British aristocrats. Instead it came from the incomes, pensions, and savings of thousands of ordinary British people the length and breadth of the United Kingdom. Rural vicars, country doctors, merchants, local bankers and solicitors, retired army and navy officers, pensioners, Oxbridge dons, shopkeepers, and myriad other individuals risked their money—in some cases most of their capital—on the future of Mexico. Why did they do it? Why did the four Cazenove brothers, of the banking family, or the three Twining brothers of tea fame, or Dr. Peter Roget, author of the famous Thesaurus that most of us still use to this day, or Henry Fox Talbot, photography pioneer, or George Sandeman, of the port dynasty, or Benjamin Disraeli, young, ambitious, and anxious to make his fortune—why did such people choose Mexico for their investments? What did they know about the country, its history, economy, politics, and prospects? What was it that persuaded them to risk their money in this way? What was it about Mexico that convinced them that their money was safe and that it was a good place to invest? Where did they get their information, living in many cases hundreds of miles away from London where news of politics and economic opportunities was almost always first known. Who were all the investors? It is with these and many other related questions that this book is concerned.

For three centuries, British politicians and merchants had looked with envious eyes at what they supposed were the vast natural resources of Latin America, untapped and largely unexploited by the Spanish colonial masters of the southern continent. Nothing had excited their interest more than the legendary silver mines of Mexico and Peru and there was a widespread belief that if the continent could be freed from Spanish rule, a whole new world of unlimited trade opportunities would be opened up to British initiative, expertise, and capital. It was not just the commercial community across Britain that increasingly looked to the Americas. Scientists of every discipline were always anxious to have access to hitherto unexplored regions and the vast range of natural fauna, both known and unknown, that were presumed to be there. Antiquarians and those interested in the ancient world wanted to travel to the New World in search of knowledge of the mysterious Inca, Mayan, and Aztec peoples and their civilizations, conquered and almost obliterated by the Spanish invaders of the sixteenth century.

From the sixteenth to the eighteenth centuries, Spanish colonial policy had been extremely effective in excluding other European nations from their American empire. Barriers to direct trade had been rigorously enforced and the New World had been effectively closed to non-Spanish people and, to some degree, ideas. Then, with the advent of the Bourbon dynasty in Spain in 1700, the situation began to change, albeit very slowly at first, and by the end of the eighteenth century, force of circumstance as well as policy had brought about some relaxation of the restrictive policy. International affairs, the pressures applied by other nations, military and diplomatic, and Spain's own weakness meant that it was no longer possible to keep foreigners out of the imperial possessions. Events in the first decade of the nineteenth century, most notably the invasion of Spain by Napoleon in 1808 and the abdication of the Spanish Bourbon royal family in favor of Napoleon's brother Joseph, caused an uncontrollable chain of events in the Iberian peninsula that led to the outbreak of wars of emancipation across the Spanish American empire. In 1810, from Chile in the south to Mexico in the north, revolution spread across the continent as region after region declared independence from Spanish rule. Ten years of war followed as Spain attempted in vain to suppress the movements for liberation and in the case of Mexico, it was not until 27 September 1821 that the final victory of the insurgent campaign was achieved when Agustín de Iturbide led his victorious army into Mexico City.

Britain's role in these momentous events is beyond the scope of this study but it may be said with certainty that the British people had watched and welcomed enthusiastically the slow but inexorable destruction of the Spanish empire in America. Although coverage was at best patchy and intermittent, news of the battles, victories, and defeats of Simón Bolívar and the other great liberators had filtered back to Britain via the newspapers and journals of the day. The British public knew what was happening, if only in general terms, and they were aware that with the defeat of Napoleon in Europe, and the collapse of the Spanish empire in America, the strategic map of the world was being radically altered. A limited amount of trade had been conducted for some years with Mexico, or New Spain as it was known under Spanish rule.[3] But now with the emancipation of the colony and the removal of Spain's restrictions on foreign trade, major opportunities for trade and investment were opened up just at the moment when Britain's own industrial and commercial development required new markets for its rapidly expanding manufacturing production.

The opportunity for investment in Mexico, therefore, first became fully and freely available in 1821 when the new independent government declared an end to Spanish trade restrictions and opened the country's ports to international commerce. But what did the British public know about what was now the former colony of New Spain? What information was available about its economy, industry, agriculture, population, transport networks, and markets? There was no shortage of books about the country and its history.[4] The story of the conquest and the heroic deeds of Hernán Cortés and the conquistadors in their campaigns to defeat the Aztecs and their emperor Montezuma had been the subject of numerous books, plays, songs, poems, and operas. Translations of works on the conquest by Spaniards and other Europeans, for example, Solís, Bernal Díaz, Clavigero, were readily available. There were also travel books the most notable of which was by the itinerant Dominican priest, Thomas Gage. First published in 1648, his account of his travels through Mexico and the Americas attracted great attention and new editions appeared regularly every few years until 1711. There were also occasionally works that attempted to describe the country's resources and potential. One example was the book published in 1741 by Scottish born John Campbell. Its title indicates its range: *A Concise History of the Spanish America: containing a succinct relation of the discovery and settlement of its several colonies: a circumstantial detail of their respective situation, extent, commodities, trade etc. And a full account of the commerce with Old Spain by the Galleons, Flota, etc. As also of the Contraband trade with the English, Dutch, French, Danes, and Portuguese*. Campbell's work is filled with statistics of Spanish America's trade and the value of its agricultural and mining production, mostly "collected from Spanish writers." Referring specifically to Mexico, it was, he wrote, "rich in all sorts of desirable commodities, gold, silver and precious stones not excepted" (p. 108). The climate was "pleasant and temperate," the soil "remarkably fruitful," and "all the necessaries of life are cheap." As for Mexico City, "in point of regularity, it is the best built city perhaps in the Universe."[5]

A few years after Campbell's book, Richard Rolt produced his *New and Accurate History of South America.* Published in 1756, it was dedicated to "the Lords Commissioners of Trade and Plantations, and to the merchants of Great Britain."[6] Rolt emphasized the wealth of America and especially its mineral resources that were producing for Spain, he claimed, 4,000,000 pieces of eight in gold per year and 30,000,000 pieces of eight in silver. In addition, he stressed, there were riches in precious stones including pearls, emeralds, turquoises, and amethysts. Fortunes were also to be made in logwoods, cochineal, indigo, and hides. Two years later, Juan Ulloa's very popular travelogue, *Voyage to South America*, appeared in English and there were other travel books and works about the Spanish empire.

Another very influential book published in 1777 was William Robertson's *History of America.* Like Campbell and Rolt before him, Robertson emphasized the wealth of Spain's empire and particularly "the vast quantities of gold and silver" to be found there. He calculated that Spain had received from its colonies £4,000,000 a year since 1492, making a total over the previous 283 years of £1,132,000,000. Adding to that figure the value of smuggled gold and silver, the total rose to "at least two thousand millions of pounds sterling." Adding a word of caution, however, Robertson also noted that the profitability of mining in America was "much decreased" but "the fascination continues." In his other chapters, he described the commercial system, the institutions of government, the agricultural economy of the continent, and the "destructive commerce" imposed by Spain's monopoly.[7]

Robertson's two-volume work was a best seller of its time, running into many more editions. Copies quickly found their way into libraries all over Britain and also into the private collections of such eminent people as the Scottish economist Adam Smith; the poet Robert Southey; and the historian Edward Gibbon. The success of his work stimulated increased interest in Spanish America and in the years between 1780 and 1810, there were many new publications about New Spain. Some were travelers' accounts like that of Zebulon Pike, who described Mexico City as "one of the greatest cities in the world," with "beauty, riches, magnificence and splendour."[8] Others were geography books in which there were brief descriptions of New Spain and Spanish possessions in America. Not all were particularly accurate. In a popular school textbook edited by Reverend Goldsmith, a pseudonym for Sir Richard Phillips, Mexico is described as "the capital of all Spanish America," but like every book of its type, it also emphasizes the wealth of the continent: "South America is best known for its gold and silver mines which have been wonderfully productive for Spain."[9] A more scholarly and better informed work was published by William Walton in 1810. Again the title reveals the range of topics it contains: *Present state of the Spanish colonies:*

including a particular report of Hispaniola, or the Spanish part of Santo Domingo: with a general survey of the settlements on the South continent of America, as relates to history, trade, population, customs, manners, etc. with a concise statement of the sentiments of the people on their relative situation to the mother country etc. by William Walton Jun., Secretary to the expedition which captured the city of Santo Domingo from the French; and resident British agent there.[10]

Unusually, Walton did not devote much space to the Mexican mining industry, preferring to concentrate on population, climate and agriculture. He did, however, make one passing comment that seems to have been part of the Mexican myth prevalent in Britain for many years. In the travel accounts published by Hakluyt in the late sixteenth century, there is the largely fictional story told by Marcos de Niza, who claimed to have discovered the city of Cíbola in northern New Spain where "they use vessels of gold and silver for they have no other metal."[11] Walton made the same point. Mentioning the abundance of precious metals, he added that "many of their household utensils are made of gold and silver."[12] A few years later, at the height of the British investment boom in Spanish America, it was stated in the House of Commons that "silver is in such plenty that the meanest utensils are made of that metal."[13]

Many more books, plays, and a new opera about Mexico—first performed at the King's Theatre in the Haymarket, London in 1803—were published in the first decade of the nineteenth century but none compared in range or quality of data to what was found in four volumes that first appeared in English in 1811. British knowledge of Mexico was transformed by the publication of the work of the famous German scientist, geographer, and naturalist, Alexander von Humboldt. His *Political Essay on the Kingdom of New Spain* became the basic source for anyone interested in Mexico.

Accounts of Humboldt's travels and career are well known and there is no need to repeat the detail here. With permission from the Spanish government, he and his companion, Aimé Bonpland, spent four years from 1799–1804 traveling around the southern regions, followed by one year in Mexico. On his return to Europe, Humboldt began to write up and publish the results of his research and it is no exaggeration to state that his books were enthusiastically received in political, commercial, and scientific communities. His work on Mexico provided an immense amount of information and statistical data on the country's geology, soil, climate, population, agriculture, industry, and trade together with his observations on the indigenous cultures of Mexico, It also offered detailed information on the silver mining industry, its scale and production methods as well as what seemed to be authentic statistics on the value of silver produced during the years of colonial rule. In short, Humboldt gave his British readers what appeared for the first time to be an accurate picture of Mexico, its economy, and its people. He also, it may be noted, depicted a land of unlimited opportunity, forecasting that "Mexico seems destined to exercise some day a great influence on the political events that agitate both continents."[14]

Humboldt's work on Mexico was widely reviewed and praised in journals throughout the United Kingdom. All manner of extracts, selections, and digests promptly appeared for those who could not afford the £3 13s. 6d. that the four volumes cost. For at least the next thirty years, no book about Mexico, and especially its mining industry, was to be found without some direct or indirect reference to Humboldt, even if it was only to correct some of his mistakes.

By December 1821, therefore, when the news of the consolidation of Mexico's independence was first received in Britain, there was a substantial corpus of written material about the country, its people, its economy, and above all, its mining industry. Yet *The Times* noted on 2 November that, "there is no part of the world concerning which so little information is possessed as Mexico, where the old Spanish system of exclusion has been in full force during the struggles for independence." The explanation for this apparent paradox lies within official opinion in British government circles. The United States recognized Mexican independence in 1822 but the British government was much more circumspect. Although strongly pressed by merchants in Manchester, Liverpool, Glasgow, Dublin, London, and elsewhere to recognize the new nation as soon as possible, the government hesitated. There were several reasons for the delay, among which were doubts of the quality of the information known about Mexico, even that provided by Humboldt. It was felt necessary that special and at times secret envoys should be sent to assess the new situation and it took time for their reports to be received. Within the cabinet there were strong monarchists and supporters of legitimacy. They argued that to recognize regimes created by revolution against legitimate monarchs like Ferdinand VII, the king of Spain, would establish a dangerous precedent. More progressive voices in Whitehall, notably the Foreign Secretary, George Canning, while understanding the strategic importance for world trade and politics that the emancipation of Spanish America implied, also wanted to be reassured that stable governments were being created across the continent. Mexico's early experiences were not encouraging in this respect. The liberator, Agustín de Iturbide, promptly established an imperial regime with himself at its head as the Emperor Agustín I. Opposition from republican groups was soon evident and Iturbide was forced to abdicate in March 1823 after just ten months in power. His imperial creation was dismantled and after some domestic conflict over the kind of republican government to be formed—federalist or centralist—agreement was reached in 1823 that Mexico was to be a federal republic known as the United States of Mexico.

By the beginning of 1824, therefore, people in Britain who had money to invest had the reassurance that the new nation had a republican government that promised to be stable and progressive. The opportunities to take advantage of this situation came early in 1824. There were two principal ways to invest in Mexico without actually traveling to the country. One was to purchase Mexican government bonds that were first offered for sale in the London market in 1824 and 1825. Following the end of the Napoleonic era at Waterloo in 1815, European reconstruction was funded by borrowed money, and with the success of the postwar French reparations loan (arranged by Barings in 1815) numerous other foreign issues were released in the London market. Russia, Austria, Spain, Portugal, Denmark, Greece, and Naples were among the nations whose securities were offered for sale. The emancipation of Spanish America also brought several of the new nations to London in search of capital. They included Chile, Colombia, Peru, Guatemala, Buenos Aires, and Mexico, all of which offered bonds for sale. In the case of Mexico, there were two issues, each with a total face value of £3,200,000. Thousands of British investors chose to buy these Mexican bonds in the expectation of receiving a high and secure level of return on their money.

I examined the whole issue of the bonds and bondholders in a major work published a few years ago and they are not my concern in this study.[15] This present monograph is about the alternative form of investment offered to the British public. This was by means of joint stock companies in which shares were traded on the London Stock Exchange. While there were investors throughout the United Kingdom who opted to put their money into the bond issues, mistakenly believing that they were a secure investment, there were also many others who preferred to take a riskier path by buying shares in various new companies that were established in London in 1824 and 1825 to invest in Mexico. There were myriad reasons why so many people became shareholders or "proprietors" as they were called, but one of the main ones was the presentation or image of Mexico at the time that the shares were offered for sale in 1824 and 1825. The first chapter of this study examines how Mexico was presented in Britain in those years. While Humboldt remained a key source of information, cited by every author, there were new more up-to-date books and articles, press coverage, personal experiences of Mexico, private and diplomatic correspondence, and most important, Britain's first ever exhibition of Mexican life and culture. All these sources of information were available to potential investors to help, or persuade, them to invest in Mexico.

In the second chapter, the formation of the joint stock companies and the market bubble that followed are explained. The companies formed to invest in the Mexican economy and especially in its silver mining industry were inevitably caught up in the bubble and the eventual market collapse that followed. While some failed to survive, for the most part the Mexican mining companies escaped the panic. Chastened by the experience, their directors, managers, and shareholders retained their conviction that investments in Mexico would prove to be rewarding. In chapter 3, the different types of companies that were formed are explained. Mining was certainly the favored activity, but in the early years it was widely believed that there were also good opportunities for investors in agriculture, pearl fishing, colonization, and many other areas. Chapter 4 deals with all the mining companies formed in 1824 and 1825 to work Mexican mines. Some were ephemeral but most were successfully established and had their shares listed on the Stock Exchange. Details are given of those behind the companies, the promoters, directors, bankers, and others involved. Large amounts of capital were quickly raised from shareholders and after a rush to acquire the most promising mines, miners were recruited from Cornwall and elsewhere, machinery and equipment were ordered, and ships were contracted to carry everything across the Atlantic to mining areas in the north and south of Mexico. The next chapter, 5, looks at what happened to the mining companies that began working in Mexico in the 1820s. Their finances are explained as are the problems they met in the fields of labor and technology. The British experience in the silver and to some extent gold mining industry was not what had initially been expected and some of the difficulties encountered with Mexican political and cultural customs are described. None of the companies made any significant profits and only one survived into the 1850s. The final section in this chapter traces the end of each company.

This first phase of British investment in Mexico in the mid-1820s has been referred to by many authors.[16] Economic historians, and those who have examined British relations with Latin America, have usually assumed that the difficulties encountered by the mining companies meant that no further British investment in Mexico took place until the final quarter of the nineteenth century. In fact, the investments of the 1820s were only the first phase. From 1830 to 1850, several more joint stock companies were set up in London to invest in Mexico's mining industry. Although on a reduced scale, shares were offered in the market and more investors were attracted by the promise of rich profits from bonanzas in gold and silver. With only one or two minor exceptions, the companies analyzed in chapter 6 have not been noted by previous authors. The Peñoles Gold Mining Company, the Zacatecas Mining Company, the Mexican and South American Company, and the Guadalupe y Calvo Company have escaped attention and yet they were all owned and managed by British investors who provided the capital and the expertise required for their operations. Even more unknown is the remarkable history of the El Bote Mining Company. A privately owned mining venture that began in 1850 in Zacatecas, it proved to be a highly profitable investment for a British and an Italian family for almost one hundred years. While other foreign businesses failed to survive the vicissitudes of Mexico's unstable and often violent history throughout the nineteenth and early twentieth centuries, the El Bote Company not only prospered but was able to pay its fortunate owners regular and handsome dividends until the 1930s.

Most emphasis in the economic historiography has naturally been on British investments in the mining industry but as we discuss in chapter 7, there were many other areas in which British capital found a place in Mexico's economy. British entrepreneurs were involved in the foundation and management of several of Mexico's mints and were therefore responsible for much of the country's coinage that was manufactured on machines and equipment bought from Matthew Boulton's famous Soho factory in Birmingham. Britons were to be found in the textile and paper manufacturing industries, in cochineal and timber concerns, as well as more basic services such as plumbing and building contractors. Although in relatively small numbers, British entrepreneurs were active in most parts of Mexico, setting up manufacturing businesses or mercantile companies that carried out much of Mexico's international trade.

This book is largely about British capital and the means whereby it was invested in Mexico in the years from 1824 to 1860. But it is also about people and, in particular, those thousands of Britons who chose to invest their money in the companies created to exploit Mexico's natural resources. In every chapter, therefore, details are given, and in some cases biographical data, on the people involved in the activities both in Britain and in Mexico. Most important in this respect is chapter 8, in which the ordinary shareholders in all the British companies have been identified. More than 1,000 names have been traced from a myriad of sources and they reveal that interest in Mexico was widespread across all regions of Britain and across most middle and upper socioeconomic groups. It would be an exaggeration to say that everybody who had the means wanted to invest in Mexico, but it is not far from the truth to assert that a majority believed that the fabled land of Montezuma and the Aztecs was destined to be rich and prosperous and a good place to invest.

Finally, in place of the usual conclusion, I have included a Miscellany that incorporates a variety of aspects of the British experience in Mexico and seeks to explain some of the paradoxes that characterize it. For example, the joint stock companies failed to make profits and the shareholders mostly lost their money. Yet, some of the British managers of the companies became millionaires. Notwithstanding the problems and loss of so much British capital that was invested in Mexico in the 1820s, there was never any shortage in later years of people still willing to risk their savings in new ventures. Their faith in the country's potential may have wavered but it was never extinguished. The revolutionary violence of Mexico, its wars against France and the United States, and its chronic political instability with governments and constitutions changing with often bewildering frequency, affected British attitudes but only marginally. As one British resident observed, while playing cricket on Sunday afternoons in the shadow of Chapultepec castle, on more than one occasion gunfire was heard in the surrounding hills, but the matches were never abandoned.

NOTES

1. J. Vincent, *An Intelligent Person's Guide to History* (London: Duckworth, 1995), 1.

2. There are several ways of computing relative values. See www.measuringworth.com/ppoweruk/ (accessed August 15, 2010). The Mexican silver peso and the U.S. dollar were of equal value in the first half of the nineteenth century. There were approximately 5 pesos or dollars to the British pound sterling.

3. G. Jiménez Codinach, *La Gran Bretaña y la Independencia de México, 1808–1821* (Mexico: Fondo de Cultura Económica, 1991) has much useful information about early contacts between Britain and Mexico.

4. For a thorough survey of early works about colonial Spanish America, see C. Steele, *English Interpreters of the Iberian New World from Purchas to Stevens: A Bibliographical Study, 1603–1726* (Oxford, UK: Dolphin Book Company, 1975).

5. J. Campbell, *A Concise History* (London: John Stagg, 1741), 108–12.

6. R. Rolt, *A New and Accurate History of South America* (London: printed for T. Gardner, 1756).

7. W. Robertson, *The History of America* (London: W. Strahan, T. Cahill, and J. Balfour, 1777), 1:388–400.

8. E. Coues, ed. *The Expeditions of Zebulon Montgomery Pike to Headwaters of the Mississippi River, through Louisiana Territory, and in New Spain, during the Years 1805-6-7* (London: Simpkin, Marshall, and Co., 3 vols., 1895), 2:721. The first edition was published in the United States in 1810 and in London in 1811.

9. J. Goldsmith, *An Easy Grammar of Geography: Intended as a Companion and Introduction to the Geography on a Popular Plan for Schools and Young Persons* (London: Richard Phillips, 1805), 60–62.

10. Walton's book was published in 1810 in two volumes; London: Longman, Hurst, Rees, Orme, and Brown.

11. R. Hakluyt, *The Principal Navigations, Voyages, Traffiques, and Discoveries of the English Nation* (Glasgow: Maclehose, 1903–1905), 4:142.

12. Walton, *Present State of the Spanish Colonies*, 2:300.

13. Thomas C. Hansard, *Parliamentary Papers* (1825), 1058.

14. A. von Humboldt, *Political Essay on the Kingdom of New Spain*, trans. H. C. Woodbridge (Lexington: University of Kentucky, 1957), Book 1, 61.

15. M. P. Costeloe, *Bonds and Bondholders: British Investors and Mexico's Foreign Debt, 1824–1888* (Westport, CT and London: Praeger, 2003). For the most recent study of Mexico's so-called "London debt," see R. Salvucci, *Politics, Markets, and Mexico's "London Debt," 1823–1887* (New York: Cambridge University Press, 2009).

16. The most important works are mentioned in the text and listed in the bibliography.

Chapter One

Mexico in Britain, 1824–1825: A Positive Image

rivers of silver, sands of gold[1]

If there was one particular day when the British began to take a serious interest in the people and affairs of independent Mexico, it was on Thursday, 8 April 1824, when the doors of the Egyptian Hall in Piccadilly were opened to the public. Inside the two large galleries, visitors found exhibitions of Ancient and Modern Mexico that displayed Mexican artifacts and natural fauna, some original and some copies, which had never been seen previously in Britain. There were indigenous deities, stone and pottery figures, a jade Quetzalcoatl, an obsidian mirror and a large-scale reproduction of the Aztec Calendar stone; originals and copies of Indian codices; and a model of the ancient city of Teotihuacan. Glass display cases lining the sides of the rooms contained flora and fruits (melons, avocado, tomato, banana, bread fruit, prickly pear and many more); leather goods like saddles and boots; birds such as flamingos and hummingbirds, and mineral samples, including silver ore taken from the famed mines of the Count of Regla at Real del Monte. There were wax models of Indians in their local dress and of Mexican gentlemen in full dress and on horseback. Maguey, tuna, and hundreds of other exhibits were on display and at the end of one gallery in which a panorama of Mexico City was displayed, the whole vista was completed with an Indian from Texcoco, José Cayetano Ponce de León, patiently standing in the garden of a Mexican cottage, "docile and extremely intelligent," watching the proceedings.[2] As Dorothy Wordsworth, who with friends and relatives visited the exhibition just a few days after it opened, remarked, it was all very amusing "and the live Mexican not the least interesting object."[3]

As a commercial venture, the Mexican exhibition was an undoubted success. The entrance charge was one shilling and the catalogues two shillings and sixpence, making, as *The Times* reporter complained, the "whole expense of admissions, 3s.6d."[4] Notwithstanding this by no means insignificant cost, nearly 50,000 people were said to have seen the first and subsequent versions of it.[5] The exhibition's promoter was William Bullock. He was one of the first British visitors to travel to Mexico after the country's independence in 1821. Accompanied by his son, also William, he left Portsmouth on 11 December 1822 and arrived at Veracruz at the end of February 1823. For the next six months, they traveled around the central region of the country, spending time in Mexico City, visiting some of the nearby mining areas, and, in general, looking for commercial opportunities. They met in Veracruz, Mexico City, and elsewhere, British merchants who were already in the country seeking business or setting up offices. Bullock Sr. also acquired mining rights at Temascaltepec that were later to lead to the foundation of the Mexican Mine Company. Finally, he spent much of his time collecting the artifacts and myriad items that were to form his exhibition in London.

Back in London by October 1823, Bullock began to prepare his exhibition and also to attract publicity for it. Months before it went on public display, it was described in detail and highly praised in *The Literary Gazette*, edited by Bullock's good friend, William Jerdan.[6] Two long articles emphasized the remarkable exhibits such as "a prodigious cactus" and the "elaborate panoramic drawing of the magnificent capital of New Spain, taken from the top of the cathedral." Bullock's diligence, it continued, had produced a collection with "every material that can assist the historian, the man of science, the naturalist, and the merchant." Unfortunately, the live animals he had also brought back with him including the "wild dog of Mexico" and a new species of deer had all died but, fortunately, the Indian, probably the first to visit Europe since the time of Cortés, it suggested, was alive and well.[7] More pre-exhibition publicity followed in *The Times* (12 January 1824) and it was highly praised in Ackermann's *Repository of Arts, Literature, Fashions, Manufactures etc.*:

> the eye, wandering around the room, is delighted with the richness and variety of the natural productions, the brilliancy of the plumage of the birds (the collection of humming birds is exquisite) and the beauty and resplendent hues of the fishes and marine productions.[8]

Bullock's exhibition was unquestionably a major event in British perceptions of Mexico and it attracted the attention of many different groups. The Foreign Secretary, George Canning, for example, was reputed to have expressed the hope that "the consuls he was despatching to the New World might possess half the information Bullock's Museum could afford them."[9]

Among the young diplomats, there was Henry George Ward. He had returned from his first, very short, visit to Mexico—about four weeks' duration—on 17 March 1824, and as he noted in his book, "I passed the whole of the Summer and Autumn of 1824 in England."[10] Details of his activities during those months are scant but it may be noted that he married his wife, Emily, on 8 April at St. George's Church, Hanover Square, just a few hundred yards away from the Egyptian Hall in Piccadilly, and coincidentally the same day on which Bullock's exhibition opened. While we have no evidence, it seems not unreasonable to suppose that having just returned from Mexico, Ward may have taken his bride to see the show.

Ornithologists like William Swainson—said to be "at that time one of the most prolific and widely read ornithological writers in England"—were especially pleased by what they found in the exhibition.[11] Bullock lent his collection of stuffed birds to him and he prepared a descriptive synopsis published in June 1827 in *The Philosophical Magazine and Annals* under the title "A synopsis of the birds discovered in Mexico by W. Bullock, F.L.S. and H.S. and Mr. William Bullock, Jr."[12] Swainson bought some of the birds from Bullock and he noted that William Bullock Jr. was sending him examples from Mexico.[13] It is worth noting that Swainson also remarked that he received information from "my English ornithological friends in Mexico, many of whom are miners."

Other scientists were equally enthusiastic.[14] Sir Evarard Hume, in an article in *The Zoological Journal* on the "Mexican Proteus," noted that he had obtained several specimens from Bullock "brought back from a lake three miles from the City of Mexico where they are so abundant in the month of June, as to form a principal part of the food of the peasantry."[15] Conchologists were likewise grateful to him, acknowledging the source of their information as "a specimen brought from Mexico by Mr. Bullock."[16] Writing to Swainson on 8 February 1824, two months before the exhibition opened, the lawyer and naturalist, William John Broderip, exclaimed, "I have been let behind the curtain of Bullock's; a most extraordinary man is he. In about two months he will have ready one of the most interesting exhibitions ever seen in London."[17] Botanists were interested in the plants and seeds Bullock had brought back with him and Swainson concluded, "the intercourse which recent political events have opened between Mexico and Great Britain, promises to be no less interesting to zoological science than important to the commercial prosperity of both nations."[18]

Many other notable figures saw and were influenced by the exhibition, and it is likely that the young Benjamin Disraeli was among them. As we will see later, he was very interested in the former Spanish American colonies and was an active speculator in Mexican and other mining stocks. It is worth noting that in his first novel *Vivian Grey*, published in 1826, he includes a satirical chapter (Chapter VII) about the new Latin American birds that everyone was talking about. The Dr. Spix he describes as "a most excellent traveller" and an expert on birds, was Jean de Spix, whose book *Serpentum Brasiliensium Species Novae* had been published in French in 1824.[19]

Bullock's contribution to British knowledge of Mexico was not confined to the exhibition. Within a couple of weeks of his return from Mexico, he wrote to the publisher, John Murray, to explore the possibility of a book about his travels. He wrote that he had just returned home bringing with him "a very interesting assembly of Natural History in all its departments. Two fine cabinets of Minerology, Antique Idols, Manuscripts, Maps etc. and drawings of the remarkable scenery and principal cities that have never yet been published." He also had, he added, "numerous notes on the country, its productions, manufactures, trade. If you think it a subject likely to interest the public in a publication I shall be glad to see you upon it."[20] Murray clearly responded positively to the enquiry because in June 1824, he published Bullock's account of his travels. The book was an immediate success with all 1,500 copies of the first edition—double the usual 750—taken up by booksellers on the first day, and further editions promptly issued in 1825.[21] The work gave a glowing picture not only of Mexico as a country for tourists but also as a place of unrivaled opportunities for commerce and investment. "The scenery," he wrote, "was inexpressibly grand"; "apples, pears and peaches almost obstructed our way"; "fields of wheat and maize as rich as any part of Devonshire"; and a local population "courteous, gentle and unoffending." Mexico offered a vast market, he emphasized, for British manufactures and products like English beer and porter that were in great demand. There was almost no domestic production of cutlery and general hardware, optical instruments, barometers, or cast iron goods, and European textiles like muslins and calicoes were much preferred over the coarse locally produced woollens and cottons. He urged manufacturers to send their samples to the many British merchants and "several respectable English houses" which were already there, adding a warning that import duties and other levies were high. Care in the choice of products, he advised, was also necessary: "grates, fire-irons, fenders, and hearth rugs have been sent to people who never saw a fire-place or a chimney."[22]

It was not just manufacturers who were encouraged by Bullock. Gardeners who subscribed to the first volume of a new annual publication that appeared in 1826—*The Gardener's Magazine*—found several items based on or taken from his book.[23] One concerned the cultivation of the agave cactus and the production of pulque in Mexico. Another discussed pig breeding or "Feeding Swine in Mexico" and a third briefly mentioned the *chinampas* or floating gardens that Bullock had described. Even more to the point was the following paraphrase of Bullock's words:

> Emigration. A few intelligent gardeners from the north of our island would soon acquire fortunes in the neighbourhood of Mexico, and be the means of increasing the blessings already enjoyed by the people in those delightful regions. In such a country and climate, the finest productions of every part of the globe might be produced, but the whole is left to the care of untutored Indians whose horticultural knowledge is indeed very limited.

Copies of Bullock's book soon found their way into local libraries in Leeds, Norwich, Newcastle, and elsewhere across Britain. Within months French- and Dutch-language editions had appeared and there was a second English edition in 1825. Later writers often referred to it and visitors to Mexico used it as a guidebook, sometimes with unfortunate consequences. Mark Beaufoy, who was in Mexico from 1825 to 1827, recorded a story he had heard of an English gentleman riding around Mexico City "comparing the open pages of Bullock's book, with what he saw himself." Unfortunately, so intent on the book was he that he failed to see the approach of robbers who attacked him and stole his possessions.[24]

The Mexican exhibitions at the Egyptian Hall remained open until September 1825, after which all the contents were disposed of by private sale or public auction. After the first few months, public interest waned and although it is not entirely clear when the change took place, it seems that Bullock, probably around November 1824, decided to combine the two separate exhibitions—Ancient and Modern Mexico—into one, and a new catalogue, the third he produced, was published.[25] With every intention of returning to Mexico as soon as possible, he then sold the lease of the premises to his nephew, George Lackington. Having sold all his business interests in Britain, Bullock and his family sailed again for Mexico to rejoin his son, William Jr., and to work the mine at Temascaltepec.

We will find more of Bullock and his family in other chapters of this work but before leaving him at this point, it is worth noting the three catalogues he produced, one each for the Ancient and Modern Exhibitions of 1824, and one for the combined display of Ancient and Modern Mexico in 1825. Each was accompanied by lithographs that provided visual representations of the scenes to be found in the galleries of the Egyptian Hall. Bullock used these catalogues not just to introduce the various items on display but also to emphasize Mexico's economic potential:

> Mexico, the unreal *El Dorado* of Elizabethan times, seems destined to become, in our day, really what it was pictured centuries ago. It has obtained a distinct political existence; its long-rivetted chains, and its long-endured slumbers, have been broken; it starts and struggles at first—there is a soreness of the fetter, and the confusion of the waking; but there is also the riches and the strength of nature, which need but to be cherished in order to convert a poor powerless province into a wealthy and mighty empire.

Mexico, he continued, was "that incalculably extensive mart to the operations of our commerce" and "one vast field for the exercise of British capital, machinery, and industry."[26]

Bullock's exhibitions, catalogues, and book were fundamental parts of the presentation of Mexico's economic and cultural attractions to potential investors in the country and to the British public in general. Their undoubted popularity was also due, Bullock claimed, to their novelty. There was, he noted, "a rising interest to this portion of the world" and yet, "I am acquainted with no book of travels by an Englishman in that country."[27] He was, however, not the first Englishman to go to Mexico since 1821—he himself names many of the merchants who were already there when he arrived—and he was also, strictly speaking, incorrect in his claim that there were no previous English travelers' accounts of the country, its population, and its economy. In April 1824, some months before the appearance of Bullock's work, Captain Basil Hall had published his *Extracts from a journal written on the coasts of Chile, Peru and Mexico in the years 1820, 1821 and 1822.* Written by an already established travel writer, Hall's book was very well received, meriting three editions in 1824 alone. No doubt to the delight of the publisher, Archibald Constable and Co. of Edinburgh, who used extracts from it in its advertising for the book, it received a favorable recommendation in a speech made in the House of Commons on 15 June. Introducing a petition from London merchants urging the government to recognize the independence of the new American states, Sir James Macintosh said, "If I am further called upon to describe the value of a full intercourse with South America, I shall refer to a very valuable publication, which I suppose every man has read, and which I know every man ought to read, I mean Capt. Basil Hall's work on South America."[28] In fact, notwithstanding this endorsement, as far as Mexico was concerned, there was not a great deal of new information in the work to attract the attention of potential investors. Hall's visit was restricted to the Pacific coast, where he arrived at Acapulco in March 1822. For the next three months, he traveled up the coast, meeting English merchants already at Tepic and San Blas and in general becoming acquainted with recent political events, notably the final defeat of the Spaniards and the rise to power of Agustín de Iturbide with his Plan of Iguala. He described these and later events in his book and citing Humboldt among other sources, offered some suggestions regarding possible trade opportunities. Then, after about three months, he left Mexico for Rio de Janeiro, a voyage, he tells us, of 8,000 miles, which took twelve weeks and five days at sea.

Bullock, therefore, was not entirely accurate in his claim to be the first British travel writer to have been in Mexico since independence, but he was correct in his statement that there were few up-to-date works available. There were other books about the country written since 1821, but for the most part they tended to be histories of events. One example was the work of the

American merchant, William Davis Robinson. His *Memoirs of the Revolution of Mexico*, first published in Philadelphia in 1820 and in London the next year, offered the British public one of the earliest accounts of the war of independence. Although the narrative is mostly given over to the Mina expedition and to Robinson's own claims against the Spanish government, the preface to the London edition included some observations that Bullock and others would certainly have noticed.[29] For example, statements such as "the extent, the fertility, the rich and varied produce of these possessions, give to them an importance which it is impossible to over-rate." Many people, it continues, had taken a keen interest in the struggles for emancipation in Spanish America, and "the merchant and manufacturer saw, in regions hitherto almost closed to commercial enterprise, a boundless scene for his operations, and inexhaustible mines of wealth." The "inestimable resources" of Mexico would ensure its liberty and "civilization and intelligence will follow in the train of commerce."[30]

Like Bullock's and Hall's books, Robinson's work was promptly to be found on library shelves around Britain, for example at the Lyceum Library in Liverpool, and the Literary and Philosophical Society of Newcastle upon Tyne. Another book published before Bullock, but which had significantly less impact, was Nicholas Mill's *The History of Mexico*, which appeared in May 1824. Notwithstanding the title, this was one, if not the first, of what were to be many publications written specifically to encourage British investment, especially in the Mexican mining industry. Referring to "the intense interest that has been excited in the public mind towards that beautiful and fertile, but hitherto neglected, portion of the globe," Mill offers his readers comprehensive details of Mexico's geographical features, trade and commerce, imports and exports, mineral production, natural fauna, population, customs, religion, climate, and disease. Mexico, he predicted, was likely to enjoy "a long period of sunshine and prosperity."[31]

The appearance of Mill's book with its emphasis on mining and commercial opportunities may well have been linked to the simultaneous formation of companies planning to invest in Mexico. Another publication that appeared a few weeks later left no doubt that the intention was to promote interest in the companies. The third edition of the English translation of Humboldt's seminal multi-volume work on New Spain appeared in March 1824, priced at £3 13s. 6d. For those who could not afford the four volumes on offer and perhaps did not have the time or energy to read them page by page, John Taylor, a mining engineer and one of the main promoters of British investment in the Mexican silver mining industry, offered the alternative of what he called *Selections from the works of the Baron de Humboldt, relating to the Climate, Inhabitants, Productions, and Mines of Mexico*. This very much abbreviated survey of Humboldt's statistics and economic data was advertised on 22 April (*The Times*) at the much more affordable price of 12 shillings. The advertisement was clear:

> This work is intended to communicate information on those interesting subjects, and particularly to those who may be investing capital in the mines of New Spain, or to others who may contemplate residence in the country.

The above publications, of course, were not the only sources of information about Mexico available between 1821 and the appearance of the joint stock companies in 1824. The growing number of those interested in the ancient civilizations of America would have welcomed the appearance of a work describing the "ruins of an ancient city discovered near Palenque in the Kingdom of Guatemala."[32] There were new editions of works on the sixteenth-century discoveries and conquests such as Elizabeth Helm's popular translation from the original German of J. H. Campe's *Cortez, or the Conquest of Mexico, as related by a father to his children, and designed for the Instruction of Youth*. This, together with works on Columbus and the conquest of Peru, were advertised by the publisher in January 1824 as appropriate Christmas and New Year gifts. For more up-to-date descriptions of the country and its economy, largely drawn from Humboldt, there were geography books, maps, encyclopedias and miscellaneous publications. Compilers and editors updated their works to reflect the growing public interest in Mexico and the other new South American states. One example was *The Geography, History and Statistics of America and the West Indies*, by H. C. Carey and J. Lea. Advertising a new edition in May 1824, the publishers, Sherwood, Jones, and Co., were careful to stress that it now came "with considerable additions, relative to the new states of South America, Mexico, the Havannah etc." More publicity for Mexico followed with book reviews in periodicals of the time. Hall's book, for example, was reviewed in the *Edinburgh Review* (vol. LX1XX) and the *Literary Gazette* (17, 24 April, 1 May 1824). Bullock's book was also reviewed very favorably and long extracts published in the latter (19, 26 June, 3 July 1824). The *Repository of Arts* included an article, together with colorful illustrations, entitled "Costumes of the Mexicans" taken from Bullock's book and another piece headed "A Spanish American dinner party," taken from Hall's description of his experiences in Tepic, Mexico.[33] The *Quarterly Review* offered a thirty-four page survey article of six books about Mexico, including Robinson's *Memoirs of the Revolution.*[34] Rather than commenting on the particular works, the writer chose to produce a synopsis of Mexican history together with a detailed survey of its population, economy, cities and, of course, its mineral output. Quoting Humboldt extensively, he gave several examples of mining bonanzas such as those of the Count of Regla, who took a net profit of more than £1,000,000 a year from his Real del Monte mines, or the Marquis del Apartado, who received £800,000 in just six months from his Sombrerete mine. Readers were assured that Mexico was certain to become a prosperous and eminent nation in the future. It had rich natural resources and was at the commercial crossroads of the world, "her Western shores at nearly the same distance from India as her Eastern shores are from Europe." Finally, *The Gentleman's Magazine* in its January 1824 issue told its readers that Mexico's mines, if neglected because of circumstances or natural causes, were still rich in silver and that British skills and equipment would overcome any problems.

Statistics of the size of editions and the number of copies sold in the 1820s are few and far between and it is difficult to estimate how many readers would have had the opportunity to read about Mexico in the 1824 edition of Humboldt's work or in those by Bullock and the others we have mentioned. According to Dr. Altick, "the average edition of a serious book was around 750 copies" but sales were obviously affected by price and as parliament was told in 1818 "books are a luxury, and the purchase of them has been confined to fewer people."[35] To some extent, the expansion of libraries in all parts of Britain offered some compensation to the reading public. London was very well served with the libraries at the British Museum and the learned societies as well as the many subscription and other lending libraries. In the country at large, there were major libraries like the Bodleian at Oxford and one contemporary estimate states that there were at least 1,000 Circulating Book Societies, each with twenty to forty subscribers who circulated books among themselves.[36] A nice example of these clubs is to be found in the correspondence of William Swainson. In a letter to him of 27 July 1824, the physician and scientist, Thomas Stewart Traill, writing from Liverpool, told him, "I have not yet read Bullock's book which is making the tour of our book club." He was sure, he added, that notwithstanding Bullock's scant knowledge of Spanish and his brief sojourn in Mexico, the book would be interesting given the author's known "industry and zeal."[37] Subscription libraries with members each contributing one or two pounds per annum were also to be found in most towns and it is largely their extant catalogues which enable us to know which books about Mexico were available outside of the metropolis.

Even the subscription libraries, however, were restricted to perhaps a couple of hundred members, although they were from the "gentry, clergy, professional men, manufacturers and well-to-do trades people" or precisely the sort of relatively affluent groups who had money to invest.[38] Given the limited accessibility to books—prices did begin to fall in late 1820s—most British people probably derived their knowledge of Mexico from the press. But again, before examining press coverage of Mexico, it is worth noting that newspapers in the 1820s, both the London dailies and the provincial weeklies, were also expensive. *The Times* and other London dailies were 7d., the equivalent according to Dr. Dawson of over £2 today.[39] Readership, however, was much higher than the circulation figures suggest—28,000 a day for the seven London morning papers, 11,000 for six evening papers and 110,000 a week for the Sunday papers, in 1829—because there were various other means whereby those who could not afford to buy could still read them.[40] The London coffee houses often subscribed to one or more and most provincial towns also had reading rooms where for a guinea a year, subscribers had a choice of papers. Here it is worth noting that the North and South American Coffee House opened on 15 September 1825 in Threadneedle Street. The proprietor, a Mr. Davis, promised to provide for his subscribers all available information about Spanish America.[41] Newspapers could also be hired, albeit illegally. "A London newsman," Dr. Altick tells us, "might make seventy or eighty lendings of *The Times* in a day, at a penny an hour."[42] The same copies would then be sent to the provinces and passed around from town to village.

Access to the press, therefore, was widespread, especially among the middle class with money to invest. To some extent, newspaper editors reflected this in their coverage of Mexico and the other Spanish American republics, emphasizing the commercial opportunities that had or would arise from the break-up of the Spanish empire and the opening of hitherto closed markets to international trade. To cite Dr. Dawson again, "Latin American items frequently occupied over 20 percent of the total columns devoted to news reports."[43] In the early years after emancipation, however, 1821, 1822, and 1823, most attention was paid to South America rather than to Mexico. The activities of Bolívar and the other liberators, and events in Buenos Aires, Colombia, Chile, and Peru, received extensive coverage. Rather surprisingly, Mexico was not especially prominent. Certainly, in the final months of 1821 when its independence was achieved with the entry of Iturbide into Mexico City on 27 September, there was noticeable press coverage. *The Times*, having commented, as we noted earlier, that "there is no part of the world concerning which so little information is possessed as Mexico," welcomed the commercial potential of the new nation, predicting great benefits to it from increased trade with Europe and with China and India. Freed from "the old Spanish system of exclusion," mining productivity would increase, agriculture would flourish, and demand for European manufactures would rise. Finally, the long-dreamed-of inter-oceanic crossing over the Mexican Isthmus would bring untold benefits, if it could be achieved.[44]

Following this initial enthusiastic attention, there was comparatively little press comment. *The Times* (18 February 1822), reporting the capture of Veracruz by the insurgents, assured its readers that "thus the revolution of Mexico is completed." Thereafter, important events such as the rise of Iturbide to imperial status were reported and British readers were kept up to date with subsequent political developments, but in the absence of regular ships to and from Mexico, information was sporadic and often gleaned from U.S. and European sources which were not always accurate. The first few months of 1823 were dominated by news of events in Spain and the French invasion led by the Duke of Angoulême that brought about the fall of the liberal government and the restoration to absolute monarchist power of Ferdinand VII. It was widely known around Europe that he had vowed to reconquer his lost American colonies but merchants and other potential investors in Mexico were reassured that such threats were entirely unrealistic. "In Mexico," *The Times* noted on 20 November 1823, "not a Spanish soldier remains."[45]

The collapse of the Iturbide empire and the formation of a federal republic were widely reported, but the most important topic in the final months of 1823 was the question of British recognition of independence. Merchants continued to petition the Foreign Office to formalize relations with the new Spanish American nations, with the appointment of consular agents as a preparation for full diplomatic recognition.[46] Canning responded with the dispatch of envoys, and in October, it was revealed that consular agents had been named for Mexico City, Veracruz, and Acapulco.[47] *The Times* (25 September 1823) took the announcement of consuls to be confirmation that full recognition was merely a matter of time and "the world will then have to consider the former Spanish colonies of South America as to all intents and purposes recognized by England in the character of independent powers." Furthermore, readers were told (19 December 1823), the Mexican leader and future president, Guadalupe Victoria, "looks with great anxiety to England, and would eagerly embrace any opportunity of establishing permanent relations with this country."

In general, therefore, the image of Mexico given in *The Times* and other London newspapers was very positive. The new nation's domestic political difficulties were reported, but they were not seen as harbingers of future instability, and potential investors had the assurance, notwithstanding the early prevarication, that the British government would recognize Mexican independence and that the threat of Spanish reconquest could be ignored. This positive image was consolidated in the early months of 1824. In January of that year, all the newspapers carried details of the first Mexican government loan floated on the London Stock Exchange. It was oversubscribed with the bonds quickly rising to a healthy premium. The quarterly payment of dividends was regularly announced on time in the newspapers, and those who had bought the bonds collected the 5 percent return on their investments. As part of the loan contract, some of the bonds were regularly bought in at market prices and long lists of the numbers of the bonds redeemed also appeared in the newspapers, giving the impression of sound financial management and responsibility on the part of the Mexican authorities. Details of the second loan—the so-called Barclay loan to be issued in January 1825—were known to the public in November and again served as further confirmation that Mexican credit in the London market was sound. *The Morning Chronicle* was enthusiastic in its praise. Mexico's potential wealth was so great that investors need have no fears. Moreover, investment in Mexican silver mines was "a good way of employing the surplus of British capital." The country with its "great resources and immense mart which her population and her wealth open to the consumption of British labour" guaranteed that Mexico "will rise into independence and greatness."[48] *The Literary Gazette* (14 August 1824) cited an unnamed correspondent in Mexico, "I have seen enough to convince me that princely fortunes might be made in cotton and indigo, without risk."

Events in Mexico were reported as usual, and outside of London the provincial press repeated the same message. *The Bristol Mirror* (9 February 1824), for example, assured its readers that "the utmost harmony prevailed." Cornish newspapers, where recruits were being solicited for new British mining companies, were similarly optimistic about Mexico's future. *The West Briton and Cornwall Advertiser* (14 May 1824) reported that "the whole of the interior is perfectly tranquil." There were more debates and merchant petitions to parliament from Leeds, London, Manchester, and elsewhere urging diplomatic recognition. On 11 June, *The Times* commented, "We hardly think that his Majesty's Ministers can hold out much longer." A few weeks later, the departure to Mexico of British diplomats was disclosed and it was again taken for granted that recognition was imminent. This impression was strengthened even more with the arrival in London in June of Mexican diplomats led by Mariano Michelena and Vicente Rocafuerte.

Diplomats were not the only Mexicans in Britain in the early 1820s. Mexicans, for example Lucas Alamán and Tomás Murphy, came to London for the specific purpose of promoting investment in their countries, and they displayed considerable initiative and skill in their cause. They were ably assisted by Mexican merchants long since established in London, such as Francisco Borja de Migoni. Even accepting that there was a well-disposed and receptive audience for their lobby among the British banking and mercantile community, the Latin Americans seem to have been easily integrated into London society, and into the corridors of power. Disraeli was careful to include in *Vivian Grey* a scene of a dinner party, held at the residency of "Old Premium," a character almost certainly based on the financier and company promoter, John Diston Powles. Among the guests present at the dinner are Latin American diplomats and "a naval officer, just hot from the mines of Mexico" and their "perpetual babble" is about "rising states" and "new loans," and "enlightened views," and "juncture of the two oceans," and "liberal principles," and "steam boats to Mexico."[49] Disraeli's fictional scene was mirrored in reality on more than one occasion. For example, the merchant banker John Hullett invited the Escandón brothers, both Mexican businessmen, to dinner at his home and noted that the main topics of the dinner conversation were "the political and economic state of Mexico and the mining enterprises."[50] Perhaps even more revealing are the remarks which John Biddulph, a wealthy banker and landowner in Herefordshire, confided to his private diary. Referring to a dinner he had recently attended, he wrote, "It was a singular circumstance to see around the same table the natives of Mexico and South America who but a few years ago were not permitted by the Spaniards to leave their own country." "All seemed men of ability," he went on, and "a very few years only will be required to make South America a much greater preponderating Power than Europe, if they all have the wisdom to keep at peace with each other."[51]

The most notable of all Mexicans in Britain in 1824 was the ex-emperor, Agustín de Iturbide. He arrived in London in January 1824 and after a few weeks passed in what seems to have been largely social engagements, he moved to Bath in March where he was soon joined by his wife and six children. Again, his activities largely seem to have been tourism and taking the waters, although he did find time to travel to Taunton in Somerset where "he intrusted [sic] two of his female children to the Catholic establishment adjoining this town, to be educated in that respectable seminary."[52] Not long afterward, he traveled to Southampton from where he left for Mexico on 11 May. His short sojourn in Britain had been reported in the press but not extensively and we do not know what, if any, contacts he had with those promoting investment in Mexico. Soon after his departure, Michael J. Quin, published an English version of his memoirs entitled: *A statement of some of the principal events in the public life of Agustín de Iturbide written by himself, with a preface by the translator, and an Appendix of documents.* In a letter to Quin, Iturbide claimed that Mexican provinces were disunited and that there was a threat to the nation's independence. His mission, he insisted, was to restore peace and if he were successful, commerce and mining would be revived.[53]

Iturbide's subsequent fate and his execution need not detain us except inasmuch as his return to Mexico did apparently cause some concern in Britain, largely because of worries that it might provoke instability and disruption to trade. *The Times* (6 August 1824) reported rumors that there was a strong pro-Iturbide feeling in Mexico, especially among the clergy and military, but on the next day, 7 August, it published a long proclamation by the Mexican government warning of the dangers to republicanism. More confusion prevailed over the next few weeks as contradictory reports of what had happened to the former emperor circulated, and it was not until late September that confirmation of his execution was published. The circumstances of that event were then given extensive coverage in the newspapers.

Another source of information about Mexico available to potential investors in much of Britain came from British merchants and others who had quickly established themselves in various parts of the new nation soon after independence in September 1821. Although exact dates of their arrival are not available, it is known that by 1826 at least fourteen British commercial establishments had opened in Mexico City.[54] Some of these had branch offices at Veracruz and at the Pacific coast ports, and they were soon presenting pleas and petitions to Britain's diplomatic representatives requesting, for example, more Royal Navy ships which they could use to send specie and bullion back to Britain.[55] Trade expanded in the early enthusiastic conviction that Mexican markets were ready and able to absorb growing quantities of British merchandise. In 1823, fifteen British ships arrived at Veracruz and Alvarado bringing 1,912 tons of manufactured goods. The next year, 1824, the numbers rose to twenty-nine ships and 3,853 tons of cargo.[56] Individual merchants and entrepreneurs also spread out across the country in search of opportunities. Among them was Messrs. Roskells of Liverpool who, according to Bullock, were the first English firm of jewelers and watchmakers to open a branch in Mexico.[57] Here we may note that Robert Roskell was Bullock's brother-in-law and one of the executors of his will.[58] Charles Whitehead was based at Campeche, where he was acting as the representative of the Liverpool business of John Forsham and Co.[59] Edward Penny from Cumbria, accompanied by a Mr. Spooner and Mr. Thomas Ferns from Manchester, went to Guadalajara in search of business. Robert Ackermann dispatched his son George to Mexico early in 1825. He took with him a large quantity of books, agricultural machinery, surgical equipment, and musical instruments with which to set up in business in Mexico City, and as a trained mining engineer, he was also to keep an eye on the mining companies in which his father was a shareholder.[60]

The letters, both personal and business, that these merchants sent back to Britain provided up-to-date information and opinion on Mexican affairs which enabled potential investors to reach decisions and assess the risks involved. A good example is a letter dated Mexico City, 24 August 1824 from J. P. Penny, brother of the Edward Penny referred to above.[61] Writing to a business colleague, Thomas Parker, who had a large textile company, Penny explained that his own affairs were going well. Business was daily improving and expected to increase in October when the rainy season had ended. "The demand for British manufactures is daily increasing," he said. He was dealing in textiles on which the profit margins were 10 percent to 15 percent, and he had recently sent more orders home. As for Mexican politics, everything was calm and nobody believed rumours that a Spanish fleet with 20,000 men was off the coast. In his first letter home from Mexico City, George Ackermann told his father that "It is impossible to conceive the business people are doing here and we will no doubt soon make immense fortunes."[62] Hullett Brothers, writing to Lucas Alamán on 28 September 1824, told him that private letters from people in Mexico to their friends in London with information about the mining companies had had the effect of raising public interest and were the cause of an increase in share prices.[63] Cornish miners who first began to travel to Mexico in 1824 to work in the silver mining industry also soon began to write home to tell of their experiences. Very few of their letters have survived, but one extant example is a letter from Captain Thomas Garby, sent from Guanajuato on 18 September 1824 to J. T. Tregelles of St. Agnes, Cornwall. Garby tells his friend of the conditions he and his fellow miners had encountered since their arrival a few weeks before. The climate, he writes, is "fine and salubrious," and "the air is scented by a profusion of the sweetest perfuming plants that can possibly be imagined." Food supplies such as beef, mutton, and lamb are "all remarkable cheap" although vegetables were scarce and expensive. On the other hand, the Mexicans, especially the poor, were "disgustingly ugly in their persons," "ignorant, superstitious, bigoted and treacherous." The priests were "a horde of unprincipled bigots," and the English had been told that it might be a good idea to pose as Roman Catholics.[64] Notwithstanding such difficulties, however, Captain Garby was reported to have concluded that "there is no doubt of the profitable manner in which the Mexican mines can be worked by British capital and miners."[65]

Personal and business letters were useful in forming opinion at a private level. There was also a substantial body of diplomatic correspondence which although in theory confidential, almost certainly provided valuable information to those with access to politicians and civil servants.[66] Contact with those with recent personal experience of Mexico was similarly helpful. Although we have no statistics of British migrants to and from Mexico in the early years of the nineteenth century and especially after independence, it is clear that many did find their way across the Atlantic. Some returned to London in 1823 or 1824 and their views were eagerly sought. For example, there was the slightly mysterious figure of William Dollar. Apparently an English merchant who had resided for several years in Panama, he arrived in Mexico in 1822. There he was soon in contact with various mine owners and quickly traveled to Britain to act as their representative with the incipient companies.[67]

There was another means whereby Mexico was brought to public attention. Perhaps surprisingly, notwithstanding the at least moral support given to them by people in Britain, the wars of emancipation and the careers of the great liberators of Spanish America seem to have attracted little attention by novelists or poets. Passing reference may have been made but no novelist appears to have thought even to write a fictionalized account of the epic struggles that took place across the continent between 1810 and 1824. In the case of Mexico, the explanation may lie in the fact that after the capture and execution of the first insurgent leaders, notably Miguel Hidalgo and José María Morelos, the war was largely conducted at the guerrilla level and no heroic figures on a par with Bolívar or San Martín emerged. Independence was achieved as the result of an obscure compromise which for some time was not fully understood in Britain, and there was no attempt to create a cult or heroic figure out of Iturbide. On the other hand, the eventual confirmation of Mexico's liberation from Spain, as we have seen, was widely publicized and welcomed throughout Britain. Bullock's exhibition, particularly of "Ancient Mexico," and the reissue of works about the discoveries and conquest renewed public interest in the epic story of the encounter between the Spanish and Aztec civilizations. In 1822, for example, Hugo Balfour (or H. J. Belfour?), who wrote under the name of St. John Dorset, composed a new work which he entitled "Montezuma; a tragedy in five acts." Dedicated to Lord Byron, and written in what Dorset hoped was an elevated verse style worthy of what he considered was an epic story, the work failed to reach the stage. It received only a modest review in *The Gentleman's Magazine*: "As a play, it is regular and well conducted, but rather too long."[68] The 142-page text was published, however, in 1822 and advertised again in *The Literary Gazette* (8 February 1823) and *The Times* (10 February 1823).[69] While it is strictly a literary work in which there is no attempt to draw the readers' attention to the wealth of Mexico, there are occasional references to "our mines of gold and silver" (speech by Montezuma, p. 33) that were at least reminders of the commercial potential of the country.

A few months later, another play was announced, described in the original playbill as "a New Grand Historical Drama (with Musick [sic.])." Written by an already very popular dramatist, James Robinson Planché, the play was entitled "Cortez: or The Conquest of Mexico." Theatergoers were promised original music by Mr. Henry Bishop, and entirely new scenery, authentic costumes, and feats of horsemanship performed on the stage by the equestrian performer and former circus artist, Mr. Andrew Ducrow. The first performance took place at the Theatre Royal, Covent Garden on 5 November 1823. In its review three days later, *The Literary Gazette* gave it a mixed notice. It was, the reviewer wrote, "a sort of musical melodrama," not altogether devoid of interest but at the same time not entitled to rank above mediocrity. The music, he continued, was good and the actors did their best, although he was not impressed with their makeup: "while Miss Love was the color of a walnut, she looked of the hue of a turnip"—Miss Love played Doña Marina. Notwithstanding this moderate review, audience reception of the play seems to have been positive because, according to one contemporary source, it ran for more than eighty nights, which certainly made it what we would call a hit.[70] After the London run ended, it moved to the provinces, opening at the Theatre-Royal, Bristol, on 5 April 1824, together with Mr. Ducrow and his horses. The local reviews were favorable. *The Bristol Gazette* (8 April 1824) had this to say:

> The new piece on Cortez, performed on Monday and Tuesday evening, was well received with great applause. The Scenery, Dresses and grand style in which the Drama is got up together with the other novelties presented . . . do great credit to our worthy and spirited manager.

After Bristol, the play was performed elsewhere in the country, at Leeds, for example, on 21 June 1824.[71]

A melodramatic and romanticized depiction of the main events and characters of the conquest, Planché's play no doubt impressed its audiences, as the contemporary reviews confirm.[72] Like Dorset's work, there are no direct references in the text to commercial opportunities or silver mines, nor would they be expected in a theatrical piece, but the image of Mexico as a land of wealth and rich natural resources is certainly confirmed in some of the poetic imagery, as this passage from Act 1, Scene 3, indicates:

> this fairy region, where spring forever reigns upon the earth, and summer wantons in the air. Whose rivers, shaming Lydian Pactolus, have waves of silver rolling over sands of gold.

Finally, we return to William Bullock and his son. In December 1825, or just a few weeks after the closure of the exhibition in the Egyptian Hall, another major exhibition about Mexico opened. John and Robert Burford were well-known artists of the time and owners of the Rotunda in Leicester Square. This circular building was used to present large-scale panoramic views of scenes painted by the Burford brothers. Using the sketches made by Bullock's son, they painted what was said to be 2,700 square feet of scenes of Mexico City, presented in a 360-degree panorama. Once again, with an entrance charge of 1 shilling, the exhibition was very popular with the public, including Reverend Skinner, who went to see it on 8 December 1826 and after spending an hour looking at it, he moved downstairs to see paintings of the Spanish capital, or as he put it, "I descended to Madrid."[73] *The Times* (13 December 1825) gave it a good review, praising the execution and skilled work of the artists, even if the city itself was not as impressive as it had once been, now spoiled by the "repulsive hue of the colours on the walls of some buildings" and "the glare of the porcelain with which some of the roofs are inlaid." A catalogue, priced at sixpence, and possibly written by Bullock, was also available.[74] It included two plates in which the buildings and other locations shown in the panorama were identified by numbers. All the major landmarks in the city were mentioned—palaces, cathedral, churches—together with the parks, the botanical garden, factories, and notable topographical features such as the lakes and the volcanoes of Popocatepetl and Ixtacihuatl. The whole was skillfully designed to present the viewer with a comprehensive picture of the city's main features. Bullock, if he was the author, used the preface of the catalogue once again to highlight the potential wealth of Mexico. In this passage, he offered a concise summary of all the pro-Mexico propaganda which he and others had presented in the years from 1821–1825:

> Possessing all the material for a most extensive exchange, Mexico offers the greatest advantages to commercial speculation: the employment of its present very numerous and unemployed population, by a combination of British talent, capital, and machinery, together with an active and unrestricted intercourse, will raise them from the lethargy and shackles in which they have been so long bound by the narrow and barbarous policy of Spain, to that rank amongst nations, which, from their character, the fertility of their soil, and the almost inexhaustible riches of their mines, they are so well calculated to maintain.

These myriad sources, from Bullock's exhibition at the Egyptian Hall, diverse books and articles, pamphlets, newspapers, theater, and private correspondence, to the Burfords' panorama, all contributed to the favorable image of Mexico that was presented to the British public in the first four years after independence. The widespread public interest, or "Mexicomania," to use Professor Leask's very pertinent word, was confirmed by another correspondent who wrote in a private letter of July 1825 that Colombia, which had earlier attracted most British attention, "is greatly thrown in the shade because of interest in Mexico owing to the greater facility of reaching the Capital as well as the greater population and richness of the Country." A lot of people, he went on, had traveled to Mexico and, "We are much better informed as to Mexico than as to Colombia."[75] All over Britain, people had taken note. To cite one example, writing from Newcastle-upon-Tyne to a colleague on 1 May 1824, George Stephenson, whose son and brother had both been offered posts with Mexican mining companies, expressed the hope that his own pioneering engineering work might yield returns "almost as good as one of the Mexican gold mines."[76] Mexicans, it was widely said, wanted British people and investment. Readers of *The Leeds Mercury* (20 March 1824), for example, were told that the British were greeted in every town and village as the "harbingers of liberty and civilization" and that an "enthusiastic feeling prevails in favour of England" (3 April 1824). *The Morning Chronicle* (21 September 1824) published a letter from a Cornishman recently arrived at the silver mining town of Real del Monte. "The people at our arrival," he said, "evinced every demonstration of joy by ringing the bells, and a deputation of the priest and principal inhabitants of the place immediately waited on us." For those back home who might be concerned about the Mexican weather, he reassured them that "here is a very fine climate, not at all hotter than in England."

Mexico, perhaps more than any of the other emerging states of Latin America, attracted the attention of British merchants, entrepreneurs, and those looking to invest in the expanding London stock market. All manner of companies were formed or projected to invest in the new republic, but their appearance coincided with a major stock market Bubble. Share prices in many companies including the Mexican and other Latin American companies reached unsustainable levels and it was not long before the Bubble burst. It is to the 1825 stock market Bubble that we now turn.

NOTES

1. The quotation is from J. R. Planché, *Cortez: Or, The Conquest of Mexico: An Historical Drama in Three Acts* (London: John Lowndes, 1823), 16. The play was first performed in London on November 5, 1823.

2. The quotation is from the catalogue of the exhibition: *Catalogue of the Exhibition, Called Modern Mexico; Containing a Panoramic View of the City, with Specimens of the Natural History of New Spain. Now Open For Public Inspection at the Egyptian Hall* (Printed for the Proprietor, London, 1824).

3. Dorothy Wordsworth to John Monkhouse, April 16, 1824, in *The Letters of William and Dorothy Wordsworth: The Later Years*, ed. E. de Selincourt (Oxford: Oxford University Press, 1939), 141.

4. *Times*, April 8, 1824.

5. *Morning Chronicle*, August 22, 1825.

6. Years later, Jerdan wrote a very flattering portrait of Bullock in W. Jerdan, *Men I Have Known* (London: George Routledge and Sons, 1866), 67–82.

7. *The Literary Gazette*, December 13, 1823; January 3, 1824; January 10, 1824.

8. Bullock had taken some "volumes of Ackermann's fashions" with him to Mexico and claimed that the designs were adopted by women at Jalapa. W. Bullock, *Six Months' Residence and Travels in Mexico* (London: John Murray, 1824), 484–85. I have relied mostly on the reprint issued by Kennikat Press Scholarly Reprints, Port Washington, 1971.

9. Jerdan, *Men I Have Known*, 76.

10. H. G. Ward, *Mexico in 1827* (London: Henry Colburn, 1828), 2:256.

11. E. Stresemann, *Ornithology from Aristotle to the Present* (Cambridge, MA: Harvard University Press, 1875), 177. Swainson offered to draw the birds in the collection in 1824, even before the exhibition was opened to the public. See the letter from Bullock to Swainson dated February 12, 1824. Linnean Society of London, Swainson Correspondence, vol. 1, Bullock letters.

12. Swainson also acknowledged "the ornithological discoveries of Mr. Bullock" in a 1828 article title "On several groups and forms in ornithology, not hitherto defined." in *The Zoological Journal*. London: W. Phillips, 1828: 3: 158–175. Later ornithologists acknowledged the debt they owed to Bullock, regretting that none of the birds he discovered had been named after him: see, for example, G. D. Rowley, ed., *Ornithological Miscellany*, vol. 2 (London: Trubner, 1876–1878).

13. On a printed page, headed "Trochilus latipennis. Broad-shafted humming bird," Swainson noted that he purchased "the female at Mr. Bullock's sale." Swainson Papers, Museum of Zoology, University of Cambridge.

14. Bullock was already a familiar figure to the scientific community, especially the naturalists, from the auction sale of his previous collections. See J. M. Sweet, "William Bullock's Collection and the University of Edinburgh, 1819," *Annals of Science* 26, no.1 (1970): 23–32.

15. *The Zoological Journal* 2 (April 1825): 127.

16. *The Zoological Journal* 1 (June 1824): 266.

17. Linnean Society of London, Swainson Correspondence, vol. 1, Broderip letters.

18. Linnean Society of London, Swainson Correspondence, vol. 1, A–C.

19. The edition of the novel I used was published in 1926 by the London publisher Peter Davies.

20. Bullock's letter to John Murray is in the John Murray Archive. A copy was kindly supplied to me by the archivist, Mrs. Virginia Murray.

21. Bullock, *Six Months' Residence and Travels in Mexico*; N. Leask, *Curiosity and the Aesthetics of Travel Writing, 1770–1820: "From an Antique Land"* (Oxford: Oxford University Press, 2002), 302.

22. The quotations from Bullock's book are from pp. 222–29, 348–65.

23. *The Gardener's Magazine*, ed. J. C. Loudon (London: Longman, Rees, Orome, Brown, and Green, 1826). The first volume appeared in the autumn of 1826. The articles referring to Bullock are on pp. 206, 447–48.

24. M. Beaufoy, *Mexican Illustrations, Founded upon Facts with Observations upon the Government and Resources of the Republic of Mexico, as They Appeared during Part of the Years 1825, 1826, and 1827* (London: Carpenter and Son, 1828), 75.

25. *A Descriptive Catalogue of the Exhibition, Entitled Ancient and Modern Mexico . . . Now Open at the Egyptian Hall* (London: Printed for the Proprietors, 1825).

26. The quotations are from pp. 3–4 of the *Catalogue of the Exhibition, Called Modern Mexico* (London, 1824). The other two catalogues were: *A Description of the Unique Exhibition Called Ancient Mexico* (London: Printed for the Proprietors, 1824); *A Descriptive Catalogue of the Exhibition Entitled Ancient and Modern Mexico* (London: Printed for the Proprietors, 1825).

27. Bullock, *Six Months' Residence and Travels in Mexico*, vi.

28. For the debate on recognition, see the *Times*, June 16, 1824.

29. The Spanish-born Francisco Javier Mina led a pro-independence group in Mexico in 1817. After several battles with Spanish loyalists, he was captured and executed in October 1817.

30. W. D. Robinson, *Memoirs of the Revolution of Mexico, with a Narrative of the Campaign of General Mina, Anecdotes of His Life, and Observations on the Practicability of Connecting the Pacific with the Atlantic Ocean, by Means of Navigable Canals* (London: Leckington, Hughes, and Co., 1821), 1:1–8.

31. N. Mill, *The History of Mexico: Also Observations as to Working the Mexican Mines by British Capital* (London: Sherwood, Jones, and Co., 1824), 282.

32. A. del Río and P. F. Cabrera, *Description of the Ruins of an Ancient City Discovered Near Palenque in the Kingdom of Guatemala in Spanish America* (London: H. Bertoud, 1824). The illustrations were drawn by the same Count de Waldeck, the French engineer and commissioner of the Tlalpujahua Company.

33. *Repository of Arts*, October 1, 1824, 234–36, 241–43.

34. *The Quarterly Review* 30 (1823–1824): 151–85.

35. R. D. Altick, *The English Common Reader: A Social History of the Mass Reading Public, 1800–1900* (Chicago: University of Chicago Press, 1963), 260–63.

36. R. Phillips, *A Million of Facts* (London: Darton and Hodge, 1832), 658.

37. Linnean Society of London, Swainson Correspondence, vol. 5, letters from T. S. Traill.

38. T. Kelly, *Early Public Libraries: A History of Public Libraries in Great Britain before 1850* (London: The Library Association, 1966), 204.

39. F. G. Dawson, *The First Latin American Debt Crisis*: *The City of London and the 1822–1825 Loan Bubble* (New Haven, CT: Yale University Press, 1990), 12.

40. Figures from Altick, *The English Common Reader*, 329.

41. Notice in the *Times*, September 16, 1825.

42. Altick, *The English Common Reader*, 322–24.

43. Dawson, *The First Latin American Debt Crisis*, 12.

44. *Times*, November 2 and December 27, 1821.

45. Spanish soldiers retained control of the fortress of Ulua, situated on an island just off Veracruz, but it was generally assumed in the press that this would soon be surrendered. It was captured by Mexican forces in 1825.

46. For one of several petitions, see the *Times*, September 12, 1823.

47. For information on Canning's difficulties, see C. Colchester, ed., *The Diary and Correspondence of Charles Abbot, Lord Colchester, Speaker of the House of Commons* (London: John Murray, London, 1861), 3:343–52.

48. *Morning Chronicle*, July 1, 2, and 3; August 14; and November 27, 1824.

49. B. Disraeli, *Vivian Grey* (London: Peter Davies, 1926), 134.

50. Hullett Bros. to L. Alamán, November 4, 1825, Alamán correspondence, Archivo General de la Nación (Mexico) (henceforth AGN), Hospital de Jesús, vol. 827.

51. Biddulph diaries, entry for December 9, 1826. There is an interesting report on what was presumably the same dinner in the *Times*, December 11, 1826. It was held at the Albion Tavern and among those present were the Mexican diplomats, Sebastián Camacho, Vicente Rocafuerte, and other members of the Mexican Legation.

52. In Taunton, Iturbide and his entourage stayed at the Castle Inn, now the Castle Hotel. *Taunton Courier*, September 22, 1824. His eldest son, Agustín, was sent to Ampleforth School, and a younger son to a school near Hampstead. *Royal Cornwall Gazette*, May 22, 1824.

53. Iturbide's letter is in the *Times*, May 18, 1824.

54. R. A. Humphreys, *British Consular Reports on the Trade and Politics of Latin America, 1824–1826* (London: Royal Historical Society, 1940), 303–14.

55. See the correspondence in the *Times*, October 19, 1826.

56. Humphreys, *British Consular Reports*, 313–14.

57. Bullock, *Six Months' Residence and Travels in Mexico*, 227.

58. Bullock died on March 7, 1849.

59. Correspondence in National Archives (UK) (henceforth NA), FO 50/8, fols. 179–93. Almost thirty years later and still in Mexico, Whitehead was appointed the agent of the British Committee of Mexican Bondholders.

60. J. Ford, *Ackermann, 1783–1982: The Business of Art* (London: Ackermann, 1983).

61. Penny's letter is in Kendal Record Office, WD/PKR/Box 1/bundle 3/15.

62. Quoted in Ford, *Ackermann, 1783–1983*, 85.

63. Alamán Correspondence, AGN, Hospital de Jesús, vol. 827.

64. Garby's letter is published in the *Royal Cornwall Gazette*, January 15, 1825.

65. *Western Luminary*, January 18, 1825. Another similarly effusive letter from Captain Garby is quoted in the *Morning Chronicle*, October 6, 1824.

66. Letters sent from Mexico to her family in Northumberland by Emily Ward, wife of Henry Ward, the British Chargé d'Affaires in Mexico, are also very informative regarding conditions in Mexico. They are in Northumberland Record Office, Swinburne (Capheaton) Estate Records, ref. ZSW/536.

67. Dollar was back in Mexico by 1828, living in poverty after losing his money in the mining speculations. He was sued for debt by Ventura Martínez and briefly jailed. Details in D. W. Walker, *Kinship, Business, and Politics: The Martínez del Río Family in Mexico, 1823–1867* (Austin: University of Texas Press, 1986), 30–32.

68. *The Gentleman's Magazine* 93 (1823): 615–16.

69. St. John Dorset, *Montezuma: A Tragedy in Five Acts, and Other Poems* (London: Rodwell and Martin, 1822).

70. The playbill issued for the Bristol performance states that the play had run for more than eighty nights in London. The original playbill is in the Bristol Record Office, Ref.9807/138. An original playbill for the Covent Garden performance is in the University of Kent, Templeman Library, Theatre Collections; Playbills, 0595182

71. Probably the most successful play of the period about the Spanish conquest was Richard Brinsley Sheridan's *Pizarro*. Sheridan's play, adapted from a German work, was first put on in 1799 and thirty thousand copies were sold in the first year. Thereafter, it seems to have been staged yearly. At Bristol's Theatre Royal, for example, it was put on nine times between 1822 and 1825.

72. There is a copy of the play in the British Library (henceforth BL) at 11778.e.3 (5).

73. Skinner journals, vol. 63, B.L. Add. 33695, fol. 184.

74. *Description of a View of the City of Mexico and Surrounding Country, Now Exhibiting in the Panorama, Leicester Square. Painted by the Proprietors, J. and R. Burford from Drawings Taken in the Summer of 1823 Brought to This Country by Mr. W. Bullock* (London: J. and C. Adlard, 1826).

75. English Papers, Suffolk Record Office, HA 157/3/118. Leask, *Curiosity and the Aesthetics of Travel Writing*, 299.

76. W. D. Skeat, *George Stephenson: The Engineer and His Letters* (London: Institute of Mechanical Engineers, 1973), 64–65.

Chapter Two

The Bubble, 1824–1826

> The year 1825 will stand like a beacon on the ocean-rocks of time, a warning to future generations—a melancholy record of human folly and credulity, almost unparalleled in the history of man.[1]

The years 1824 and 1825 have received a bad press in the annals of British history. For those who observed them at first hand, it was a time when the "sharpers of society" defrauded thousands of British middle-class families of their savings and condemned many to genteel poverty, if not to the debtor's prison.[2] "The ignorant, the excited, and the deceived," in the words of the Scottish advocate and judge, Lord Henry Cockburn (1779–1854), were swindled out of their money and "not one honest penny was made out of all the villainy and folly." Samuel Gurney (1786–1856), Quaker banker and closely involved in the events, recalled that the winter of 1825 "was a season of universal gloom in town and country."[3] Harriet Martineau (1802–1876), whose own family was badly affected financially, put it succinctly when she wrote some twenty years after the events; "There are some now of the most comfortable middle-class order, who cannot think of that year without bitter pain."[4] Walter Bagehot, writing almost half a century later in 1873, recalled "a panic so tremendous that its results are well remembered after nearly fifty years."[5]

The cause of such bad memories was a stock market Bubble characterized by apparently wild speculation and a gambling frenzy in which the English middle class throughout the country from Cornwall in the south to Northumberland in the north were afflicted by what *The Times* (5 November 1825) called the "leprous infection of avarice." The Irish and the Welsh were not immune from this "wanton demon of speculation," and as one contemporary put it, "you can have no idea of the feverish battle which all these things have created among the formerly sober citizens of Edinburgh."[6] Displaying what political economist John McCullock (1789–1864) recalled a few years afterward as "an incredible degree of ignorance and folly,"[7] respectable clergymen, doctors, solicitors, local bankers, merchants, retired military, widows, and pensioners, together with many members of the titled nobility, were drawn into "speculations of unreasoning avarice" that left behind nothing but "darkness and stifling odours."[8] Cockburn lamented, "I could not have conceived that madness could be so universal."[9]

For many contemporaries, the blame for the misery suffered by the victims of the market crash that came in the autumn of 1825 was clear. It belonged firmly at the doors of crooked lawyers, bankers, and stockbrokers, or to use Cobbett's words, the joint stock blackguards among the Jews and Quakers of Lombard Street.[10] More recent economic and financial historians have taken a less personal and broader, more pragmatic, view of both the Bubble and the subsequent market crash, criticizing the cheap credit and other lax fiscal policies adopted by the British government and the Bank of England.[11] Following the end of the Napoleonic war era in 1815 and after some initial problems caused partly by the government's deflationary policy, the British economy by 1820 had regained its expansionary momentum. Industrial production increased and exports rose substantially while excellent harvests reduced the need for expensive imports of wheat. Food prices fell, leaving consumers with more money to spend.[12] With abundant and increasing capital in circulation, much in the form of small denomination banknotes issued by hundreds of country banks, the government chose to take advantage of the favorable circumstances, abolishing some taxes and reducing its long-term debt. Interest rates on the "Funds," as they were known, were reduced. In 1822, for example, the "Navy Five per Cents" were reduced to 4 percent. Two years later, the 1789 "Four per Cents" were exchanged for 3½ percent stock. Above all, while the nominal rate remained unchanged, the real yield on the Three per Cent Consols, traditional home for investors looking for a secure place for their savings, fell from 5 percent in 1816 to 3.3 percent in 1824.

Faced with the general trend of declining rates, investors looked for alternatives "and were receptive, by 1824, to virtually anything that came on the market."[13] Several possibilities presented themselves. Some speculated in commodities like cotton, wool, silk, copper, and other minerals, the prices of which rose rapidly in 1824. For those unwilling to take risks in the always hazardous commodities market, there was another potentially lucrative investment in the form of a relatively new type of government security. At the same time that the British government's need for money declined, the demands of other nations increased and the age of foreign loans began. Governments across Europe required loans and they came to London, already the world center of international finance, to raise them. The Latin American nations were not far behind, offering bonds worth over £22,000,000 for sale in the London market for which, taking into account the discounted prices, thousands of what seem to have been largely British investors paid approximately £17,500,000.

Most of the Latin American issues were quickly oversubscribed and there were several reasons for investors' enthusiasm. First, as government-backed stocks, they appeared secure. Second, dividend or annual interest rates of 5 percent or 6 percent, but in real terms much higher because of discounted issue prices, were very attractive, especially in comparison to the falling rate of British government securities. Furthermore, there had been, and still was, significant public sympathy in Britain for the emancipation struggles of the Latin American countries against what many saw as the centuries' old tyranny of Spain. The threat of reconquest of his lost American empire by Spain's absolute monarch, Ferdinand VII, provided another reason for hispanophobic British to help Simón Bolívar and the other great liberators in their struggles for freedom from colonial rule. The profit motive was also a powerful influence. In short, as one nineteenth-century historian put it, "It was impossible to resist the combined attractions of an appeal which promised liberty to a continent, and 20 percent to an investor."[14]

Foreign government bonds were not the only option open to investors seeking to improve the returns on their money. Starting rather slowly in the early years of the century, partly because of legal problems (see below), joint stock companies began to be formed for various infrastructure projects. Most of these were for canals with the others including docks, water works, bridges, gas installations, roads, and finally, no less than thirty Assurance Companies. These new enterprises offered shares to the public and it is noticeable that their promoters and directors included a number of politicians, including Members of Parliament, and many prominent bankers and other luminaries from London's financial and mercantile community—men such as the financier and bullion dealer, Isaac Lyon Goldsmid; the bankers Sir J. W. Lubbock, Charles Bosanquet, and Sir John Perring; merchants Hieronimus Burmester and David Barclay; and future banker, Henry Cazenove. Equally, if not even more prominent, were titled aristocrats. Norwich Union Fire and Life Insurance Office, for example, had for its patrons no less than three Dukes, four Earls, four Viscounts, and other titled people, together with a Member of Parliament, a Director of the Bank of England, and the Reverend G. H. Baird, Principal of the University of Edinburgh.[15] These early joint stock companies were to be an important model and precedent for those formed in later years. Moreover, in many instances, the same men who appeared as directors of the early companies reappeared in those formed in 1824 and 1825.

According to one contemporary estimate, prior to 1824 there were 156 joint stock companies representing a combined capital of £47,936,486 13s.6d. and 764,534¼ shares. Many had paid a dividend or rate of interest higher than could be obtained from alternative forms of investment. This success probably contributed to the continued interest in that form of raising capital and from the early months of 1824 onwards, there was a rapid acceleration in the formation of new companies. Here it may be noted that none of the pre-1824 companies involved Latin America but in January–February 1824, the first three companies designed to exploit silver mining in Mexico were announced. Thereafter, month by month, more and more companies were formed until on just two days in January 1825, no less than thirty-five new ones were advertised in the press. We do not know exactly how many were eventually formed but historians seem to agree that the calculations made by the contemporary participant, and it may be acknowledged, diligent observer, Henry English, are the most authoritative. He was a London stockbroker and, perhaps to assist his own dealings, he began to note details of the new companies and to collect copies of each prospectus issued by them separately or published in the press. The result of his activities was several publications, starting in October 1825 with a pamphlet which he described as "a general guide to the companies formed for working foreign mines." This includes the prospectus of each mining company, the amount of capital, number of shares, names of the directors and a brief indication of the fluctuations in their share prices. The next year, 1826, a similar publication dealt with companies formed to work British mines and then in 1827, he produced his "complete view" of the joint stock companies formed in 1824 and 1825. In later years, English maintained his interest in the mining companies in two journals he edited, namely *The Quarterly Mining Review* (1830–1834) and *The Mining Journal*.[16]

All of English's publications are used extensively by economic and financial historians but it is his 1827 survey which provides most of the statistics. As they are readily available elsewhere, only a brief summary is given here. According to his information, 624 companies were formed or announced in 1824–1825 with a total capitalization of £372,173,000 and almost six million shares on offer. The companies covered a wide range of activities including mining (74); gas (29); insurance (20); investment (28); canal and railway (54); trading (11); building (26); provision (230); miscellaneous (292).[17] As English himself recognizes, however, many of these companies were what *John Bull* (6 February 1825) labeled "ephemeral follies," and perhaps intended by their promoters more to amuse than to attract investors. There was, for example, the Economic Meat Company; the Royal Patent Purifying Sea Water Company; the Cigar Club; the Metropolitan Bath Company, the London Umbrella Company; and the Equitable Tea Agency Company. As one Member of Parliament put it, gentlemen were well advised "to keep their money against a rainy day rather than invest in companies to bring salt water from Bognor, air from Heaven, and blasts from Hell."[18] Others were generally recognized at the time as being intentionally fraudulent inasmuch as they had no real industrial or commercial prospects and were formed simply to enable their promoters to profit by manipulating share prices in the market.[19] One example was The London Friendly Economical Society, which promised investors a return of 500–600 percent on their money. Among the more notorious frauds was the imaginary kingdom of Poyais, land of the Poyers, in Honduras. Skillfully promoted by a supreme "sharper," Sir Gregor MacGregor, as an earthly paradise with a climate so benign and disease-free that life to an old age was assured, with cathedrals, banks, and an opera house, naïve investors and migrants were deluded into buying shares or even migrating to what was in reality the mosquito-infested and hostile Indian territory of Central America.[20]

It was in January–February 1825 that the Bubble in the stock market reached its peak with share prices, especially in some of the new mining companies, reaching large premiums. With prices rising daily, investors scrambled to obtain shares which could be bought for a small deposit, and professional speculators skilled in the ways of the market bought and sold on a daily, if not hourly basis, making, it is said, huge profits. The buoyant market continued at a less feverish pace until April when the first signs of the decline were seen. Thereafter, prices in all the new companies fell and those investors who had bought their shares with borrowed money in the hope of quick profits were caught in a classic bear market. Then some companies, especially those intended as long-term enterprises, called on their shareholders to pay further installments on their shares. With share prices declining, many investors rather than risk more money, chose to sell, even at a heavy loss, and the market was even further depressed. Others hastened to withdraw funds from their banks in London or the country. In short, there was a run on the banks and they in turn sought funds from the Bank of England whose bullion reserves were rapidly depleted to such an extent that it came "within an ace of stopping payment."[21] In early December 1825, the major London banks of Wentwort; Chaloner, and Richworth; and Pole, Thornton, and Co. failed and dozens of their associated country banks were also forced to suspend payments and close their doors to their depositors. Lord Greville noted in his memoirs: "The state of the City, and the terror of all the Bankers and Merchants, as well as of all the owners of property, is not to be conceived but by those who witnessed it."[22] Seventy-three banks in England failed and three in Scotland, and in the next few months, hundreds of bankruptcies among businesses of all sorts were declared. Between 13 October and 7 November 1826, there were 636 petitions for bankruptcy.[23] Of the 624 companies listed by English only about a fifth, or 127, were still in existence in 1827. As *The Times* (21 April 1826) put it, "This humbug is now dropping to pieces: the officers, the Directors, are taking to boats and small craft on board in order to escape: whilst the crew of shareholders are left to perish with the cracking timbers."

Within the English-compiled statistics are those companies formed to invest in the newly independent Spanish American republics and Brazil. Using his figures, supplemented by other sources, at least sixty-one companies were planned, covering all the main Spanish American countries from Chile in the south to Mexico in the north and including the islands such as Haiti, Cuba, and the Philippines. Myriad industrial, agricultural, commercial, and transport projects were announced. The largest category involved mining with twenty-seven companies in Bolivia, Brazil, Chile, Colombia, Mexico, Peru, and Argentina. There were several colonization schemes allied to investments in agricultural developments in Colombia, Mexico, and Argentina. Indigo production was envisaged in Honduras and Guatemala and general trading companies were formed, or announced, for Haiti, Cuba, Panama, and the Philippines. Even more popular were ambitious plans to bring British technological expertise, mainly a diving bell, to pearl fisheries along the Pacific coast of the continent. Other precious stones were not neglected. Prospective investors in the South American Gem Company, for example, announced in March 1825, were told of the potential wealth in diamonds, rubies, sapphires, topaz, emeralds, and amethyst to be found in abundance, it was claimed, in Brazil and other regions of the continent. Finally, there were at least four schemes to introduce steam navigation between Europe and America such as, for example, the Havana and Hayti [sic] Steam and Sailing Packet Company.

These Spanish American and Brazilian projects shared the same characteristics evident in the whole array of companies. Some did not go beyond the planning stage, for example, those promising the new steam-powered navigation lines; others issued a prospectus and sold a few shares before being abandoned as unviable; and some, especially the mining companies, reached what might be described as an operational stage. Their shares were placed in the market and the large capital sums raised were spent on equipment ranging from steam engines for the mines of Peru, Mexico, and elsewhere, to ships to transport colonists to Colombia and Argentina. Again, with the exception of some of the mining enterprises, none of the companies survived more than a couple of years and many of those who invested in them lost much, if not all, of their money.

The country that attracted the largest and most varied group of the new companies was Mexico but before examining these, it may be helpful to explain some of the procedures and more general characteristics of joint stock company formation. The first controversial feature, attracting much attention and comment at the time, was their legality. According to one interpretation of existing English law, the formation of companies or corporate bodies with multiple shareholders and transferable shares was prohibited unless authorized directly by Royal Charter or an Act of Parliament. This prohibition had its origins in the Bubble Act of 1720 that had been passed because of the notorious events associated with the so-called South Sea Bubble.[24] Hundreds of bogus companies had been formed and thousands of people tricked into parting with their money for worthless shares. For several decades of the eighteenth century, company formation had been severely restricted as a result but gradually applications to Parliament proved to be more successful and more than one hundred canal acts had been passed by 1800. Other types of company, such as the British Plate Glass Company, incorporated in 1773, had also appeared. In the first two decades of the nineteenth century, as we have seen earlier, a large number of new companies were permitted and it soon began to be argued that the 1720 Bubble Act or at least its effects had lapsed or were no longer applicable. Here is not the place to go into the complexities of the law because, as was said in the House of Commons on 29 March 1825, nobody really knew what it was. Faced with the legal confusion and the *fait accompli* of the hundreds of companies formed in the preceding months, Parliament repealed the 1720 Bubble Act and although the Lord Chancellor promised to bring in new legislation to regulate joint stock companies, nothing was in fact brought forward. As Professor Hunt expressed it, "Confused litigation and the ghosts of 1720 harmed the course of the unincorporated joint-stock company until the middle of the century."[25]

Company promoters, therefore, applied to Parliament for a charter of incorporation. Such was the demand—in the 1825 session alone, some 438 petitions were presented and 286 private acts passed—that the process was protracted.[26] The presence of many Members of Parliament among the company directors may have facilitated it for some companies but not, it appears, to any great extent. Anxious to capitalize on the booming demand for shares, those behind the companies went ahead before any parliamentary or royal sanction was given.

The first stage was the publication of a prospectus, and those advertising Latin American projects offer good examples. Essentially publicity or advertising sheets, they outlined the main aims of the proposed venture and most included exaggerated claims designed to attract the unwary investor. The Anglo-Peruvian Mining Association, to give just one example, stated that in the Peruvian mining areas, gold, silver, and quicksilver were abundant; and there were "lumps of pure gold weighing from 2 to 50 pounds" lying around untouched by local inhabitants. A few years previously, it continued, a single miner had found a lump of metal worth 3,000,000 dollars. The other mining companies made similar claims. Some were careful to use the name of Alexander von Humboldt to give credibility to their statements emphasizing that their alleged facts and figures regarding the wealth or potential of their mines were based on the distinguished traveler's published works. All emphasized that with British management, capital, and technological expertise, especially the application of steam power, fortunes would be quickly made and shareholders would receive handsome dividends. In several cases, it was stated that company agents were already in America busily negotiating contracts with local mine owners who were only too pleased to receive British assistance.

The colonization companies, and those promising to develop agriculture in Brazil, Colombia, and elsewhere, painted alluring pictures of idyllic landscapes where every sort of crop was or could be cultivated with little effort. The Río de la Plata Agricultural Association declared its aim to be to offer a new and better life to unemployed agricultural laborers of Great Britain and Ireland, enabling them to establish themselves in the Argentine as independent farmers. It would pay the transport costs of "a certain number of young farming men, of good character, if healthy and strong, with their wives, if married, and not exceeding one child with each couple (unless at their own cost)." The Haytien Trading Company's prospectus began with these words: "The rising greatness of Hayti, and the well-known stability of its government" and continued with claims that British manufactures would find a ready market.

Several stressed that they had connections and approval from important Latin American politicians and diplomats. The Brazilian Agricultural and Jewel Company claimed the direct support of the Brazilian government for its plans and other Brazilian companies claimed direct grants from the Emperor of Brazil. The Colombian Agricultural Association gave a well-publicized dinner at the Star and Garter, Richmond Hill, on 19 June 1825 at which the principal guest was the Colombian diplomat and Minister to Britain, Manuel José Hurtado, who was already said to have given his advice and protection to the South American Association for Agricultural and other objects.[27] Not to be outdone, the Río de la Plata Agricultural Association sponsored a farewell dinner at the Ship Tavern, Greenwich on 16 August 1825 for the departing Buenos Aires diplomat and future president, Bernardo Rivadavia.[28] Mexican and other Latin American notables in London at the time were similarly used to give respectability to the companies.

Some, but not all, the prospectuses included the names of the directors and auditors of the company, together with names and addresses of its bankers and solicitors. Details of the capital sum to be raised by the sale of shares, voting rights of shareholders, and various other proposed regulations were also given. The shares could be bought in small installments, often starting at £5 with the remainder payable in installments from time to time "as the concerns of the Company may require it, at the call of the Directors." These future payments were henceforth known as "calls" and potential shareholders were assured that thirty days' notice would always be given before a call for payments was made. Failure to make payments resulted in the forfeiture of the shares. Directors were usually elected for five years in the first instance, after which three retired by rotation unless re-elected. Each director was normally required to hold a minimum of twenty-five shares and to be an auditor required at least twenty shares. Shareholders were entitled to vote at General Meetings according to the number of their shares: for example, ten shares gave one vote, twenty, two votes; fifty, three votes; and one hundred, four votes with no shareholder having more than four votes.

Most prospectuses referred to the controversial question of personal liability. Again without going into the legal complexities, the issue revolved around the question of the personal liability of shareholders or partners in an enterprise. Hitherto, those who had an investment or share in a business faced unlimited liability for the debts of that business on the basis that those who take the profits must also pay the losses, if necessary, from all of their private assets. In short, unlimited liability was the "normal manner of doing business."[29] The possibility of losing all one's assets was obviously a major deterrent to potential investors in new enterprises and the company usually, therefore, promised the following:

> A clause is to form a component part of all Contracts, engagements, or obligations of every kind into which the Directors may enter, stipulating that no share-holder is to be responsible for more than the amount of his or her share or shares remaining unpaid.[30]

In effect, limited liability was thus promised to investors but doubts were soon cast on the legality of the promise. Further paragraphs in the prospectuses concerned dates of annual general meetings, trustees, disposal of shares, and dividends.

There was frequent reference to a "Deed of Association" or "Deed of Settlement." All the companies prepared similar legal documents that were said to be in modern terms (1936), the equivalent of a memorandum and articles of association by which they became "common law companies."[31] Every shareholder was required to sign the Deed within a specified time or risk the forfeiture of any installments paid on shares. The newspapers of the day in 1824 and 1825 are filled with announcements by company secretaries informing shareholders that the Deeds were ready in the company offices awaiting signatures. Given the importance and number of these Deeds, it is perhaps surprising that only one original appears to have survived.[32] This is the Deed of Association of what became known as The Mexican Company (details of this company are given below). It is a long, densely written thirty-three pages, much of it in legal jargon, of all aspects of the proposed association. The statements made in the prospectus regarding, for example, capital, shares, objectives, names of directors and other employees, are confirmed. The 162 sections also set out the rules and regulations; names, duties, and powers of the directors; the financial controls and procedures; shares and shareholders; bankers, trustees, and solicitors; meetings; "salaries and compensation of the Directors, Trustees, Auditors, Solicitors and Secretary, and of all Commissioners and Agents, Officers, Bookkeepers, Servants, and Workmen"; in short, a whole range of subjects concerned with the future operations of the Company.

NOTES

1. G. H., *The American Mines: Shewing* [sic] *Their Importance in a National Point of View With the Progress and Present Position of the Real del Monte Company* (London: Effingham Wilson, 1834), 1.
2. H. Martineau, *A History of the Thirty Years' Peace, 1816–1846*. 1st ed., 3 vols. (London, 1849–1850. Reprint, Shannon: Irish University Press, 1971), 2:1.
3. H. R. Geldart, *Memorials of Samuel Gurney* (London: W. and F. G. Cash, 1857), 29.
4. H. Martineau, *A History*, 2:21.
5. W. Bagehot, *Lombard Street: A Description of the Money Market* (London: Henry S. King and Co., 1873), 178–79.
6. Quotations from A. Romney (pseud.?), *Three Letters on the Speculative Schemes of the Present Times, and the Projected Banks. Addressed to His Cousin Richard, in the Country* (Edinburgh: Bell and Bradfute, 1825), 3, 7. See also R. C. Michie, *Money, Mania, and Markets: Investment, Company Formation, and the Stock Exchange in Nineteenth-Century Scotland* (Edinburgh: Donald, c.1981).
7. J. B. McCullock, *Dictionary of Commerce and Commercial Navigation* (London: Longman, Rees, Orme, Brown, and Green, 1834), 801–04.
8. Martineau, *A History*, 2:20–21.
9. H. Cockburn, *Memorials of His Time* (Edinburgh: Adam and Charles Black, 1856), 432–33.
10. *Cobbett's Weekly Register* (London: J. M. Cobbett, February 12, 1825), 415–18.
11. See, for example, the paper by Larry Neal, "The Financial Crisis of 1825 and the Restructuring of the British Financial System," prepared for the 22nd Annual Economic Policy Conference at the Federal Reserve Bank of St. Louis, MO, October 16–17, 1997, www.research.stlouisfed.org/publications/review/98/05/9805/n.pdf (accessed August 24, 2010).
12. Michie, *Money, Mania, and Markets*, 27.
13. A. D. Gayer, W. W. Rostow, and A. J. Schwartz, eds., *The Growth and Fluctuation of the British Economy, 1790–1850* (Oxford: Oxford University Press, 1953), 1:185.
14. S. Walpole, *A History of England* (London: Longmans, Green, and Co., 1878–1886), 2:122.
15. For the names of the directors of several of the companies established prior to 1824, I used the 1819 London Post Office Directory.
16. H. English, *A General Guide to the Companies Formed for Working Foreign Mines* (London: Boosey and Sons, 1825); *A Compendium of Useful Information Relating to the Companies Formed for Working British Mines* (London: Boosey and Sons, 1826); *A Complete View of the Joint Stock Companies Formed during 1824 and 1825* (London: Boosey and Sons, 1827).
17. Figures from Gayer, Rostow, and Schwartz, eds., *The Growth and Fluctuation of the British Economy*, 1:412–14.
18. Hansard, 12:612, cited in B. C. Hunt, *The Development of the Business Corporation in England, 1800–1867* (Cambridge, MA: Harvard University Press, 1936), 37.
19. Figures from Gayer, Rostow, and Schwartz, eds., *The Growth and Fluctuation of the British Economy, 1790–1850*, 1:412–14.

20. There are several works on the Poyais fraud. Among the most useful is Richard T. Gregg, *Gregor MacGregor, Cazique of Poyais, 1786–1845* (London: Scripophily Library, International Bond and Share Society, 1999).

21. Bagehot, *Lombard Street: A Description of the Money Market* (London: Henry S. King and Co., 1873), 29.

22. L. Strachey and R. Fulford, *The Greville Memoirs, 1814–1860* (London: Macmillan, 1938), 154.

23. *Times*, November 9, 1826.

24. For an entertaining account of the South Sea Bubble, see C. Mackay, *Memoirs of Extraordinary Popular Delusions* (London: Richard Bentley, 1841), 1:74–103.

25. Hunt, *The Development of the Business Corporation*, 21.

26. Figures from Martineau, *A History*, 2:8.

27. *Times*, June 20, 1825.

28. Ibid., August 16, 1825.

29. B. Weiss, *The Hell of the English: Bankruptcy and the Victorian Novel* (London and Ontario: Associated University Presses, 1986), 137.

30. Clause from the prospectus of the Anglo-Mexican Mining Association.

31. W. F. Crick and J. E. Wadsworth, *A Hundred Years of Joint Stock Banking* (London: Hodder and Stoughton, 1936), 10, 16.

32. The deed is in the Bancroft Library, University of California. The text of some of the Deeds of Association is given in *The Quarterly Mining Review*, 4 vols. (London: T. and T. Boosey, 1830–1837).

Chapter Three

The Mexican Companies

> The wary Mexican seizes the prey and laughs at the eccentricities of 'enlightened' Europe.[1]

Mexico was the most popular of the new Latin American nations as far as merchants and entrepreneurs were concerned and its popularity was reflected in the number of companies planned or formed. Some of these did not get past the stage of initial enquiries and perhaps were no more than personal daydreams. For example, one project was to use a diving bell to raise treasure from shipwrecks in the Gulf of Mexico. Another proposed to plant one million coffee trees on 500 acres acquired in Mexico, if an investor with £10,000 could be found. An annual income of £10,000 was expected "when the trees arrive at full bearing."[2] William Elwand asked permission to set up a business in Mexico to manufacture glass and cotton fabrics for which he would import machinery from Britain. John Biddulph recorded in his diary that he had been asked for his opinion on a Philippines Company. "The plan," he went on, "is to buy the sovereignty of the isles from Spain who will sell the same for £300,000."[3] Several applications were made to the Mexican authorities for permission to develop whale fisheries, probably reflecting Humboldt's descriptions of the abundance of whales in the west coast regions of the country.[4]

Another popular idea that did not get beyond the planning stage was for some form of inter-oceanic communication across southern Mexico.[5] A group of very prominent London bankers and merchants, including Barings; John Irving; Pascoe Grenfell; David Barcla; Herring and Co.; P. Green and Co.; Herring, Graham, and Powles; and several others, proposed to form a company that would construct a navigable canal or a railroad ("*camino público de hierro*") across the Isthmus of Tehuantepec. They claimed that

Humboldt had promised to help with advice and that engineers who had built the Caledonian canal in Scotland would be employed in the construction. Contract negotiations were started in Mexico in 1825—the company wanted one-third of all profits for thirty years—but no agreement was concluded.[6] The idea seems to have been dropped for the time being but it was not abandoned. A book on the subject also appeared in 1825, offering what was described as "a succinct view and analysis on the practicability of joining the Atlantic and Pacific oceans, by a ship canal across the isthmus of America."[7]

Some ideas reached the stage of a prospectus but seem to have gone no further. There was, for example, a "Prospectus of a loan for the purpose of establishing a British Commercial Company trading from Liverpool to the republics of Mexico and Peru."[8] Another was for a Mexican Iron Works Company with an initial capital of £40,000 divided into 2,000 £20 shares. Although printed, this prospectus contained no names of directors or other details.[9] Another printed prospectus without names of the promoters was for the British and Foreign Timber and Wood Association. This proposed to import timber from around the world but especially the mahoganies and "valuable woods of New Spain and Honduras" that were in short supply and increasingly expensive.[10] Some schemes were advertised in the press but then disappeared. One example was an advertisement in *The Times* (11 May 1826) for an "Anglo-Mexican Loan." This stated that the proposed capital of £175,000 divided into £35 shares would be used to set up a company to trade with Mexico. Subscribers were promised interest-bearing bonds until profits, expected to exceed 60 percent, were enough to pay dividends. Shares could be bought from Mr. R. Piercey, 4 Angel Court, Throgmorton Street. Finally, there was the American Smelting Association. Its capital was to be £500,000 in £100 shares. It was to make money by employing new methods to work ores that had not responded to the amalgamation process used in the Mexican and South American silver mining industry.[11]

One area that did attract much interest was colonization and land development. Unlike many of the noted London banks of the time, Rothschilds and Barings had tended to stay clear of the companies formed to invest in Latin America, at least publicly. In private, the position seems to have been rather different. The involvement of Barings in the Tehuantepec canal project was noted earlier and as we will see later, Alexander Baring privately invested £5,000 of his bank's funds in one of the mining companies. The Mexican Minister Plenipotenciary in London, Mariano Michelena, reported to his government in March 1825 that Alexander Baring was very anxious to establish communications and trade with Mexico. For that reason he had sent his son, Francis, to visit the country.[12] There, in addition to acquiring estates near Mexico City, Francis signed contracts to buy an estate of some nine million acres in northern Mexico. We know little about the reasons for this purchase or intended use of the land but it involved Barings in several years of litigation, and a loss of at least £40,000. Alexander Baring disapproved of his son's actions and did his best to withdraw from the purchase. His intrigues were successful when the Mexican congress decided that foreign ownership of real estate was not permitted.[13] Allegations were later made that a number of congressional deputies had been bribed by Alexander Baring to pass the law banning foreign ownership because its approval meant that the land contracts signed by his son were invalid.[14]

There were other land and colonization schemes. In the first three years after independence, 1821–1824, successive Mexican administrations had debated at length the best means of promoting colonization by European immigrants. Several laws were passed, promising land grants, tax exemptions, and general assistance.[15] These laws were published in the British press and they stimulated all manner of schemes from many individuals. Robert Torrens, for example, then Member of Parliament for Ipswich, on behalf of unnamed backers, wrote to Vicente Rocafuerte with "Proposals for settling a Scotch and English colony in the Territory of New California." He wanted Mexico to pay the travel costs of the migrants and provide them with 3 million acres of land, agricultural tools, and provisions for one year. The organizers of the scheme also wanted £100,000 a year in Mexican bonds to pay their costs. They would provide 5,000 families chosen from "Scotch settlers selected for their sobriety and industry."[16] Mariano Michelena reported to his government another approach that promised to provide 12,000 Scottish and Irish families, and one colonel John Grant claimed that he represented numerous families in the Scottish highlands who were interested in emigrating to Mexico.[17]

None of these various approaches made much progress but there was one scheme that reached the company formation stage. This was the Mexican Colonization, Agricultural and General Trading Company, announced in *The Times*, 25 March 1825. Its two promoters were Thomas Roper Curzon and Jacob Henry Deacon. They approached Michelena offering to take 400 families to northern Mexico where they would develop wine and olive production. It seems that agreement was reached and in the advertisement of 25 March, it was stated that lands in Upper and Lower California, New Mexico, and Texas had been provisionally assigned to the company for the purposes of colonization, agriculture, and general trading together with other rights to pearl fisheries and mining. The capital of the company was to be £1,200,000 in 12,000 £100 shares. William Ball was appointed company secretary and applications for shares were invited to the solicitors, Ravenhill and Crook at 28 Poultry. Exactly what happened thereafter is not clear. Michelena sent the proposed contract to Mexico for ratification on 8 June 1825.[18] On 2 July, the Company's Committee of Management reported that favorable news had been received from Mexico and that it was intended to proceed with the original plan. Then, four weeks later on 4 August, another press release announced that it had been decided to await ratification from Mexico before shares were issued.[19] Nothing further was heard about the Company.

Another category of companies formed in association with Mexico involved the foundation of new shipping lines. Hitherto, there had been no regular cargo or passenger ships between Europe and the Americas. Merchants, diplomats, early tourists, and visitors had to rely on irregular voyages of sailing ships organized on an ad hoc basis by their owners. British Admiralty rules also dictated that specie or bullion could only be carried on ships sent specifically for that purpose. The ships used were often from the British West Indies station and the main beneficiaries were the captains who collected a 2½ percent commission. Cargoes of silver were usually insufficient to fill the holds and the captains often spent weeks sailing around the Caribbean looking for other cargoes. Long delays ensued and the mercantile community in London soon began to protest. One petition addressed to George Canning on 4 March 1824, for example, was signed among many others by Barings, Hullett Brothers, J. D. Powles and Co., B. A. Goldsmidt and Co., Green and Hartley, and the chairmen and managers of three of the recently established mining companies.[20] In response to demands from merchants, the British government agreed to the establishment of a regular monthly packet service to Mexico, usually leaving from Falmouth.[21]

Company promoters had other ideas and although the technology of steam navigation was still in its infancy, notices soon appeared promising new steam shipping lines to the American continent. One of the earliest, advertised in *The Morning Chronicle* (19 August 1825), was the London, Colombian and Mexican Steam Navigation Company. Its capital was to be £200,000 in 1,000 £200 shares, available for a deposit of £3 a share. The bankers were Barclay, Tritton, and Bevan and the office was at 38 Lothbury. Three days later, 22 August, another company was announced in *The Times.* This was the West Indian, Mexican and Colombian Steam Navigation Company. This new venture had been formed, it was said, to facilitate communication with the Americas. Applications for shares were to be made to the solicitors, Messrs. Tottle, Richardson, and Gaunt, or to the civil engineer, Mr. Wright, at 77 London Wall. Several similar companies made their appearance at around the same time and *The Times* (22 August 1825) warned prospective investors that most had no hope of commercial success and were fraudulent, their promoters hoping to profit from manipulating the share prices. Certainly, none of the promised steam shipping lines to Mexico came into operation for several more years.

The possibility of large numbers of migrants and other workers being transported across the Atlantic to work in Mexico and elsewhere in South America also attracted the attention of the rapidly expanding British insurance industry. One new company announced in February 1825 was the Crown Life Assurance Company. The prospectus stated that its main objective was to insure the lives of those going abroad. It was especially interested in those going to Mexico and South America and "the Company will be found to present peculiar advantages to persons proceeding to Mexico, South America and other parts beyond the seas."[22]

One of, if not the only joint stock company not associated with silver mining in Mexico to reach the operational stage was the General Pearl and Coral Fishery Association. According to its prospectus, issued in 1825, it had begun as a private venture but then decided that more capital was needed to add to the two ships already acquired. These vessels were to go to the Gulf of Mexico and to the Californian coasts where Humboldt had stated there were also rich pearl beds.[23] Investors were assured that the recently improved diving bell would enable divers to exploit areas so far untouched. Hence, it offered shares to the public to raise a capital of £500,000 in 15,000 £40 shares, with the usual 10 percent deposit or £4 payable in the first instance. The Board of Directors and auditors included an impressive array of prominent people from London's mercantile and political establishment. There were three Members of Parliament, two of whom were John Bent, who represented Totnes, and Richard Hart Davis, a Somerset landowner whose constituency was Bristol. The third was Thomas Wilson, a merchant by profession whose constituency was the City of London, where he was popularly known as "Buckskin Wilson." He was to be prominent within and outside of Parliament in the whole joint stock company controversy. There were two baronets—Sir Gore Ousely, former British ambassador to Persia and noted oriental scholar of his day and Sir William Adams/Rawson, who was connected with several schemes and who we will encounter again in mining industry projects. The first chairman was Thomas Hamlet, a goldsmith and jeweler of Prince's Street, Soho. Among the others, there was Thomas Meux, a brewer; and several London merchants—John Jacob, John Petty Muspratt, George Rougement, and Henry Usborne.

Although we know very little about the activities of most of the joint stock companies, in this case, rather unusually, we have two sources of information. Former Royal Navy Lieutenant R. W. H. Hardy was appointed commissioner for the company, a post he held from 7 May 1825 to 7 May 1828. On his return to Britain, Hardy published a travel book in which he recounted his experiences, including his work for the company.[24] He explains his negotiations with the Mexican government to obtain a licence to search for pearls and how he then went to Mazatlán and Guaymas on the Pacific coast. Eventually a ship, the *Wolf*, sent from Britain by the company, arrived and he left for the Californian coast. After fruitless efforts to find pearls, he was soon disillusioned. The diving bell in which he descended several times, gave him "excruciating pain that I felt in my ears and eyes" and was clearly unsuitable for oyster collection because of the uneven surface of the seabed. Only local divers, he concluded, could recover the pearls but "the whole business, from all I can learn, is a system of fraud and chicanery, consisting principally in the purchase and sale of oysters among the fishermen themselves." In short, there was no hope for "so wild an enterprise as that of the Mexican Pearl Fishery."[25]

The same conclusion regarding the company's operations was reached independently of Hardy by another observer in Mexico. This was Eustace Barron, a merchant and British Consul at Tepic. On 7 July 1826, he sent an interesting report to the British mission in Mexico City about the activities of the *Wolf* and another ship belonging to the Company, the *Bruja.*[26] The latter was a small schooner apparently brought out to Mexico by an unnamed director of the company. It had gone to the island of Tiburon where pearls were said to be abundant. No ship had gone there unescorted for many years, Barron added, because of the danger of being attacked by the local Ceri Indian tribe. The usual procedure was for three or four ships in convoy to go to Loreto in May, returning in October, each vessel taking between thirty and forty divers who descended to a depth of twelve to fourteen fathoms to search for the oysters. The diving bell that the company intended to use, however, was unsuitable and had not been properly adapted for the local conditions. Hence, Barron concluded, "the speculation" was sure to fail.

No information is available on the eventual fate of the General Pearl and Coral Company but it is unlikely to have survived much after the return of Hardy from Mexico in 1828. In the early stages, April 1825, the share price reached £38 for a share on which only the £4 deposit had been paid and speculators who bought and sold at the right time may have made money. On the whole, however, investors who, according to English, had paid £60,000 in calls on their shares lost their money. One other aspect may be noted. Several similar companies were formed in 1824 and 1825 to exploit supposedly rich pearl fisheries in Panama and Colombia. One of these was the Colombian Pearl Fishery Association. Among its promoters and financial backers was the well-known Crown goldsmiths and jewelers, Rundell, Bridge, and Rundell, of Ludgate Hill. It is not entirely clear which members of the firm became involved in the pearl fishery business. Its head was Philip Rundell, but he died in 1827, leaving, incidentally, the immense sum for the time of £1,200,000 in personal effects.[27] It seems more likely to have been the other partners, John Bridge and Edmund Waller Rundell, who was Philip's nephew. Both were already directors of one of the new mining companies, The General South American Mining Association, and Bridge was a shareholder in one of the Mexican mining companies. They had also enquired about the possibility of getting exclusive privileges from the Mexican government for a whale fishing company.[28]

Edmund Rundell was one of the founding directors of the Colombian Pearl Fishery. Its activities were in the southern continent and some documentation regarding them has survived, especially letters by him concerning the company's affairs.[29] They reveal, however, a similar story to the Mexican venture, although investors were still buying shares in 1827 or 1828—for example, one General O'Loghlin bought 100 shares. Nevertheless, the company seems to have been wound up in 1829 and once again investors lost their money.

Although a separate business from the General Pearl and Coral Company, there is an intriguing indication of links with Mexico. In January 1825, there was a launch ceremony at Sir Robert Wigram's yard at Blackwell of a vessel constructed for the Colombian Company. What *The Times*' (21 January) reporter calls "a great concourse" of people was present, not so much to witness the launch but to see the diving bell that was to be used in the pearl fisheries. The guest of honor for the occasion was the Mexican minister, Mariano Michelena, who was accompanied by his entourage, including the Legation secretary, Vicente Rocafuerte. He, we are told, with several other gentlemen present, went down in the diving bell and survived the experience. What is not clear is why what seems to have been the entire Mexican diplomatic corps was present at the launch of a ship for a Colombian-based company. All the rival companies were competing for exclusive privileges at this time and it may be that the Colombian Company was hoping to get permits to operate in Mexico. It is also noticeable that both these pearl fishery companies were promoted by London jewelers.

Finally, diverse other projects were almost certainly considered but details have been lost. The Mexican government, for example, was anxious to encourage British companies to develop the drainage works and canals in the valley of Mexico. It also wanted investment in gas lighting, roads, and transport and in a letter to the secretary of its Legation in London, Vicente Rocafuerte, it urged him to encourage all kinds of projects. It is interesting to note that in his reply on 8 October 1825, Rocafuerte advised that the earlier enthusiasm among British "capitalists" for Mexico had waned considerably. All the available capital had been used and everyone now awaited the results of investments already made. Moreover, parliamentary elections were soon due and they would absorb another £2,000,000.[30] Rocafuerte also knew, however, that one particular area of investment continued to attract widespread attention among the British public and that was the Mexican mining industry. It is to the companies formed in London to exploit Mexico's silver mines that we now turn.

NOTES

1. *Times*, August 30, 1826.
2. Advertisement in the *Morning Post*, November 6, 1824.

3. Diaries of John Biddulph, Hereford Record Office, file G2/IV/J/54—henceforth referred to as Biddulph diaries.

4. A. von Humboldt, *Political Essay on the Kingdom of New Spain*, trans. H. C. Woodbridge (Lexington: University of Kentucky, 1957), 3, 83–90.

5. The idea of an inter-oceanic canal or crossing of some sort had long interested the Spanish authorities and many others. Jeremy Bentham, for example, was concocting one such scheme in 1822–1823—for details see M. Williford, *Jeremy Bentham on Spanish America: An Account of His Letters and Proposals to the New World* (Baton Rouge: Louisiana State University Press, 1980), 87–94.

6. The various letters and documents, including the proposed contract dated April 16, 1825, are in the Archivo Genaro Estrada, Secretaría de Relaciones Exteriores, Mexico City (henceforth referred to as SRE), L-E-1510.

7. R. B. Pitman, *A Succinct View and Analysis of Authentic Information Extant in Original Works on the Practicability of Joining the Atlantic and Pacific Oceans, By a Ship Canal across the Isthmus of America* (London: J. M. Richardson, J. Hatchard, and Son, 1825).

8. This prospectus was in BL, 8223.e.10 (15). Unfortunately, it is no longer there and has not been located.

9. There is a copy in NA, E192/5.

10. This prospectus is in BL, 8223.e.10 (13).

11. Advertisement in the *Morning Chronicle*, July 15, 1825.

12. M. Michelena to Minister of Foreign Affairs, March 21, 1825, SRE, 1-14-1599.

13. Correspondence in SRE, 12-29-33, fols. 1–13, and R. Pakenham to Earl of Dudley, December 20, 1827, NA, F.O. 50/36, fols. 177–81. See also H. McKenzie Johnston, *Missions to Mexico: A Tale of British Diplomacy in the 1820s* (London: British Academic Press, 1992), 164–65.

14. The best source of information about the Barings land purchase is the court case of Kinder v. Lord Ashburton and Others, fully reported in the *Times*, November 28, 1838.

15. D. G. Berninger, *La inmigración en México (1821–1857)* (Mexico: SepSetentas, 1974).

16. Correspondence in SRE, 40-16-76.

17. Berninger, *Inmigración*, 54.

18. SRE, 5-9-8226, fol. 7.

19. *Times*, July 2, August 4, 1825.

20. NA, F.O. 50/8, fols. 139–49.

21. T. Pawlyn, *The Falmouth Packets, 1689–1851* (Truro, UK: Truran Books, 2003), 111; *Times*, August 21, 1824.

22. The prospectus is in the *Newcastle Courant*, February 5, 1825.

23. Humboldt, *Political Essay*, 3:79–90.

24. R. W. H. Hardy, *Travels in the Interior of Mexico in 1825, 1826, 1827, and 1828* (London: Henry Colburn and Richard Bentley, 1829).

25. Hardy, *Travels*, 230–449.

26. Barron's report is dated Tepic, July 7, 1826 and is in NA, F.O. 50/22, fols. 226–28.

27. Details of Philip Rundell's will are in the *Times*, February 21 and 26, 1827.

28. S. Camacho to V. Rocafuerte, Mexico, January 11, 1826, SRE, L-E-2056, fol. 4.

29. Correspondence of Messrs. Rundell and Bridge, Derbyshire Record Office, Ref. D3155/WH 2942.

30. Correspondence in SRE, 1-14-1599, fols. 4–10.

Chapter Four

Mining Companies: Phase 1 (1824–1829)

> In this age of wonders, steam engines may be applied, and do the work of wind and tide.[1]

British interest in Mexico's legendary silver production had a long history. The fabulous yields of Mexican mines such as the Valenciana at Guanajuato, and the stories of bonanzas and instant riches conferred on their lucky owners, were well known to merchants, politicians and the wider community the length and breadth of Britain. Silver, both specie and bullion, was the currency of international trade, and with the rapid growth of the British economy and expansion of trade being brought about by the industrial revolution, access and, if possible, control of the sources of silver was highly desirable, and certainly thought to be in Britain's strategic interest. Hence, the events in Spain and its American colonies in the early years of the nineteenth century, and, especially, the revolutions for independence that spread rapidly across Spanish America in 1810, had from the beginning an attentive audience throughout Britain. While the British government prevaricated over immediate recognition of the revolutionary regimes that had won control of almost the whole continent by the early 1820s, merchants and other groups were less inhibited. We do not know exactly when or who the first British visitors were to arrive in Mexico in the weeks and months following its achievement of independence in September 1821 but there was certainly a steady stream of ambitious businessmen, entrepreneurs, and adventurers anxious to exploit any opportunities they could find in the newly opened market. Among these early travelers were representatives of merchants and manufacturers of London, Manchester, Liverpool, Glasgow, and elsewhere, and above all, there were those who went to seek opportunities in the mining industry.

Mexico's silver mines were scattered across large areas of the country, mostly, but not exclusively, to the north of the capital, Mexico City. Hundreds, if not thousands, of mines had been developed during the colonial period in the northern states of Chihuahua, Durango, Guanajuato, Michoacán, San Luis Potosí, Zacatecas, and in Oaxaca to the south. They had yielded vast profits to their owners and rich tax revenues to Spain in the eighteenth and early nineteenth centuries but the whole industry had been devastated during the ten-year-long war of independence. Mines across the country had been abandoned by their owners and by 1821, after years of neglect, many were flooded and seemingly impossible to restore to production. British, and it must be said, other European nationals, notably the Germans, saw their opportunity. To their mind, there was not the slightest doubt that experienced European miners, supported by enough financial capital and, above all, the application of technological developments such as steam power, would easily be able to rescue the abandoned mines.

From the Mexican viewpoint, similar hopes were entertained. The first governments of the independent country were fully cognizant of the fact that its mineral resources were the principal source of wealth and likely to be so for years to come. They wanted, and actively sought, European investment in the mining industry, amending restrictive Spanish legislation to encourage foreign investment. But, it was not a nationalized industry. The mines belonged to private citizens. They lacked the financial resources to repair the damage to their mines and they were only too willing to discuss terms with potential British investors. Some, like the Count of Regla, appointed representatives in Europe to find investors.[2] His mines were in the Real del Monte district that had an almost legendary status in Europe as once having been the source of vast wealth for the Romero de Terreros family. There were also a number of Mexicans already in Europe, some to attend the Spanish *Cortes* or parliament in Madrid, and they took advantage of the opportunity to look for investors. The most important of these was Lucas Alamán. He was to enjoy a long and important career in Mexican politics and among his many areas of activity, both in and out of high office, was the economic development of Mexico. Born in the mining city of Guanajuato to a wealthy family of miners and landowners, he had personally witnessed the devastation brought to his local mining industry and he knew that foreign capital was essential if it were to be revived. Furthermore, he had studied mining and metallurgy at educational establishments both in Mexico and in Europe and was able to speak and write authoritatively on the subject. In Paris in the spring of 1822, where he was friendly with Humboldt, he set up the first of the new mining companies, and in July he traveled to London in search of more investors. While in Britain, he was certainly helped by Francisco de Borja Migoni, a Mexican merchant who had resided in London for several years. Another Mexican we know to have been in the metropolis at this time promoting the virtues of investment in their silver mines was Colonel John Murphy. He was a member of a prominent family of mine owners who were to arrange contracts for their mines with more than one of the joint stock companies.

Finally, there was General Arthur Wavell. A rather shadowy figure in Mexican history, Wavell had a varied career as an army officer, first in the Bengal Militia, and then in the Spanish army during the peninsular war. By 1820, he was in Chile fighting in the independence army as an infantry colonel and two years later, he traveled to Mexico. The details of how and why he went to Mexico are to say the least hazy but it is known that he managed to ingratiate himself with the Mexican leader, Agustín de Iturbide, who gave him the rank of brigadier and later full general in the Mexican army, as well as promising him a substantial land grant in Texas. According to letters Wavell wrote in 1823, when he was in London, he had been commissioned by the Mexican government to find out all he could about the most modern and up-to-date machines and equipment in use in the mining industry, and above all, to find what he called "capitalists" in Britain who would be willing to invest their money in Mexico. He returned to Mexico on 31 August 1823 but shortly before leaving, in a letter to Lucas Alamán, he told him that he would have two companions on the voyage, one a celebrated mining engineer with twenty years' experience, and the other, a representative of potential British investors. We do not know who these men were.[3]

Hundreds of Mexican mines were contracted, therefore, by the companies in negotiations conducted in London and in Mexico. The terms of the contracts varied but they conformed to the Mexican mining legislation of 1783 whereby foreign ownership of mines was not permitted. Thus, the new companies became in effect the financial backers of existing mine owners rather than owners themselves of the mines. Dr. Rankine has a succinct explanation of how this was done:

> The Companies agreed to rehabilitate the mines and, in return for this service, were to receive the profits applicable to an agreed number of the 24 shares (into which mine ownership was customarily divided) for a specified number of years—usually ranging from 10 to 16. Until the mine came into profitable production the Companies were to pay the mining proprietors an annual advance, amounting in most cases to 24,000 pesos which, under mining law, was a first charge on the profits of all the shares in the mine. Thus, until the advances had been fully paid off, there could be no division of profits.[4]

In the case of the Valenciana mine at Guanajuato, for example, the British company acquired eight shares, or *barras* as they were called in Mexico, for a term of sixteen years during which the owner was to be paid 24,000 pesos a year. In other instances, the mines or shares in them were ceded in perpetuity and in some for a relatively short period of three or four years.[5] Clauses were sometimes included that gave the Mexican owners certain rights to intervene or participate in the management of the mines. Occasionally, a *barra*, or one twenty-fourth part of a Mexican mine, was offered for sale in London by public auction.[6]

With so many mines on offer from Mexicans in London and even more so from mine owners in all the mining districts of the country, the companies competed to acquire the best available, often without prior inspection or any idea of the condition of the property. Once contracts were agreed or mines promised, the next task was to recruit personnel to send out to Mexico. Advertisements were placed in the newspapers, especially in Cornwall where it was hoped to find willing volunteers in the Cornish mining communities. Generous salaries were offered, partly payable to Cornish banks for the benefit of families left behind, and several times more than local wages. There was no shortage of recruits. Hundreds of experienced Cornish miners took what appeared to be a real opportunity. As Schwartz has put it, "This offered their families the prospect of financial stability and the added incentive of rising higher up the mining hierarchy than was possible in Cornwall."[7] Managers, accountants, medical personnel, chemists, architects, carpenters, blacksmiths, and laborers were also sought. Those with mining experience or engineering skills were particularly desirable and it is noticeable that several of the commissioners or senior staff sent to Mexico were retired army officers. Similarly, those with a good knowledge of Spanish were eagerly sought after as interpreters. James Lees is a good example. Born in Manchester, he went to Spain when he was twelve years old and lived there for several years while his father was engaged in various business pursuits. He became totally fluent in Spanish and then back in London in April 1824, he was offered a job as the interpreter/clerk for the Anglo-Mexican Company. He accepted the offer and traveled to Mexico, arriving in August 1824 with the Company's first party of miners. A couple of years later, his fluency in Spanish was so good that the Mexican authorities presumed that he must be Spanish by birth and they included him within the terms of a law by which Spanish nationals were to be expelled from the country. The Anglo-Mexican Company appealed against the decision and it was accepted eventually that he was a British citizen.[8] Another man who was offered a job, as engineer and superintendent, was Robert Stephenson. He declined the offer, partly he said, because of the disapproval of "my Wife and large family."[9] His brother, George, busily engaged at the time in developing the steam engine and the railways but always in search of orders for his engineering business in Newcastle-upon-Tyne, was also consulted.[10] His son, also Robert, did subsequently accept a post with the Colombian Mining Association and he spent three years working in South America.

As employees were being recruited, the companies began to place orders for mining equipment and above all, steam engines. Many of these orders went to Cornish foundries such as Harvey and Co. at Hayle, which built several engines for the Real del Monte Company and was in dispute over payment for them for much of 1825.[11] Another enterprising Falmouth company, Banfield and Lake, put an advertisement in *The Times* (18 April 1825) offering to supply the South American mining companies with any amount of "the best mining gunpowder." Other small businesses supplied "everything from theodolites and ropes, to miners' dags (picks), shovels, bucking irons (flat faced hammers for grinding ore), copper riddles (sieves) and kibbles (iron buckets)."[12] A wide range of other goods and materials including large quantities of quicksilver, was also needed. The cargo manifest of one of the first shipments to reach Mexico included the following:

16 boxes of instruments
16 boxes of books for record keeping
9 boxes of medicines
22 packets of paper
9 carts
12 wheelbarrows
1,000 flasks of mercury[13]

With personnel and equipment to hand, the final stage was finding ships to transport them to Mexico. Advertisements were placed in the press inviting ship owners to tender for the contracts and again there seems to have been no difficulty in quickly finding them.[14] As noted already, within just a few weeks, ships were leaving British ports en route for Mexico with company employees as their passengers and mining equipment as cargo. For example, one group of fifteen employees left Liverpool on 25 March 1824 going first to New York and then to Tampico where they arrived on 25 May. Another larger group of twenty-eight left about the same time on the *Tigre* and it arrived off Alvarado on 18 June 1824. Twenty more left from Portsmouth on 12 March 1825 and over the following months, as the new companies were organized, ships carrying passengers and equipment left from Falmouth, Swansea, and other ports. No accurate total is available of the number of these migrants who left to work in Mexico but it was certainly in the hundreds, if not more, including, it may be noted, some women and children. Given the hazards of ocean travel at the time, there was always some anxiety over the voyages and when reports of the progress of the journey were brought by returning ships that had encountered the outgoing vessels, usually in the Caribbean, they were often published in the newspapers, to the relief of families and the sponsoring company. Arrival in Mexico was an achievement in itself but in some instances, the haste of the company promoters to be the first in the field brought tragedy. Although experience and knowledge of conditions in Mexico was limited, it was widely known that yellow fever was endemic in the Veracruz region from the beginning of May to the end of October. Europeans arriving in the area during those months faced a strong risk of contracting the disease and usually it proved fatal within a few days In May 1826, the British Minister in Mexico, Henry Ward, reported on the fate of some of the mining company employees: "25 unfortunate men who came out with Captain Colquhoun perished between Alvarado and Veracruz, before the attempt to remove the steam engines last year, was given up." Arriving at the wrong time of year, he added, meant "certain death to any European." In his view, the blame for the tragedy lay firmly at the door of the company directors who knew, or should have known, of the dangers.[15]

Notwithstanding the British enthusiasm for investing in the Mexican silver mining industry, some of the companies failed to get beyond the projected stage or were of very short duration. The Biscaina and Morran Company, for example, was announced but very little information about it is available. *The Morning Chronicle* (23 February 1825) reported its formation and the imminent publication of a prospectus, although no such document appeared. It went on to say that several mines in Mexico had already been secured for it "by a Gentleman holding a high military rank in the Mexican service." It is possible that the person referred to was General Wavell. A few weeks later, James Vetch, a former Royal Engineer, and a commissioner of one of the other British mining companies already in Mexico, wrote to a colleague in London with details of the mines and land he was trying to acquire for the new company.[16] It seems probable that Vetch failed to contract the mines he wanted and there is no further mention of the company.

The Guanajuato Mining Company was similarly short-lived. It proved to be an interesting case of the fraudulent activities of the "sharpers" and it offers a good illustration of one aspect of the Bubble. The first mention of it seems to have been in *The Morning Chronicle* on 23 June 1825 when an announcement was made that because so many applications for shares in the company had been received, the books were now closed. This notice was signed by "T. Greaves, secretary pro tem, Lloyds." A week later, 30 June, a prospectus was published under the name of one Thomas Mitchell, company secretary, of 35 St. Swithin's Lane. The company, it said, intended to work silver mines in the environs of the city of Guanajuato; "the Association," it continued, "has hitherto been conducted by six Trustees in England and eight in Mexico, and the result of their labours will be seen by the Prospectus." The mines of Guanajuato were more productive even than those of the fabled Potosí in Peru. From 1796 to 1803 they had yielded almost 5,000,000 dollars a year, equal to a quarter of all the gold and silver produced in Mexico. Arrangements were being made for a Deed of Association and all the usual terms of general meetings, directors, auditors, and shareholders' voting rights were explained. A general meeting of subscribers to the shares was promised as soon as the company's agents arrived from Mexico, bringing with them the ratification of contracts signed with the Mexican government. In the meantime, shares were offered for sale. Of the 2,000 shares of £200 each, giving a capital of £400,000, 600 had already been taken in Mexico. Now, the remainder were offered for sale at a £5 deposit payable within seven days, at the bankers, Frys and Chapman.

This was a persuasive commercial for the new company and within an hour of their offer in the market according to one source, the shares were trading at a premium of between £3 and £10, that is, a £200 share on which £5 was due but not yet paid was on offer at between £8 and £15. Within a few days, however, rumors began to spread that all was not right with the new company and on 5 July its secretary, Thomas Mitchell, put out a statement denying the rumors and assuring "dissatisfied parties" that they could get their deposits on shares back from the bankers.[17] Two days later, 7 July, the Stock Exchange Committee published a warning to investors to beware of "dealing in the shares of any company, with the respectability and fair conduct of which they are not perfectly satisfied." The next day, 8 July, the solicitors, Wilks and Verbeke, wrote to the Foreign Exchange Committee of the Stock Exchange. They enclosed a copy of a letter they were sending to *The Times*. In this, they said a Mr. Stilman had approached them. He was the main promoter of the Guanajuato company and he had asked them to act as its solicitors. On questioning him, however, it had become obvious that the whole project was a fraud. Everything stated in the prospectus was a lie designed to inflate the share price. Stilman had confessed as much but insisted that by manipulating the shares, they could make £100,000 profit very quickly.[18]

Word of this letter quickly reached the market and on 11 July, the bankers, Frys and Chapman, announced that they were prepared to return all the deposits paid, amounting to £2,425, to holders of share receipts. *The Morning Chronicle* (11 July 1825), in reporting this offer, commented; "thus one out of the very numerous frauds on the public has been put a stop to." The Guanajuato Company, therefore, came to an abrupt end. The tactics adopted by its promoters had been typical of the fraudulent maneuvers employed by some of the "sharks." *The Bristol Mirror* (24 April 1824) summarized the process:

> A project is set on foot, with an air of secrecy; it is whispered about, talked about; people apply for shares, but in vain; at last the scheme is made public—then, however, the shares are at a premium of five, fifteen or even eighteen per cent. Purchases are made with avidity, the original speculators sell out and realize property—shares fall to a discount and the dupes suffer.

We have no details of who may have been duped into buying the Guanajuato company shares but given the sum that Frys and Chapman offered to return to depositors, £2,425, it seems probable that few investors lost any large sums.

More money may well have been lost in the next company to be noted. This was the Mexican Mine (or Mining) Company. Again, not much information is available but the project involved three individuals who were very prominent in the joint stock company market and in particular its connections with Mexico. The first was Sir William Adams (later Rawson) (1782–1827). Information on his life is readily available in the *Dictionary of National Biography* (henceforth *DNB*) and only a few details are required here. A surgeon by profession, he specialized in eye treatment and although his methods provoked some debate, he rose to be surgeon and oculist to the Prince Regent and several other members of the aristocracy. He was knighted in 1814 and two years before his death in 1827, he took the name of Rawson. Always a controversial and eccentric character in his medical career, his extant correspondence demonstrates that he was an ardent supporter of Latin America's emancipation and a fervent believer in its future economic prosperity. As was noted earlier, among his varied interests, he was a director of the General Pearl and Coral Fishery Association.

Adams's main Mexican interests, however, were in the silver mining industry. He held shares in at least one quoted company, and speculated in others, but his direct concern was with the Mexican Mine Company. Its origins are very obscure. Writing to Mexico from London on 16 December 1824, he said, "I have formed a mining company, capital £500,000, with power to raise more if necessary." He went on to explain that it was to be a private venture for the time being and the £1,000 shares in it would not appear in the market until "the success of the plan is proved." £20,000 had been allocated for the initial stages. He named the people behind the venture as Sir James Cockburn, Paymaster General of Marines and brother of Sir George Cockburn, a Lord of the Admiralty and related to the royal family; the banker, Sir John Lubbock; James Campbell, director of the Bank of England, and rumored millionaire. "These," he went on, "with other City bigwigs, are to form the Committee of Management" but, "your humble servant plays a very leading fiddle in all these concoctions. In fact, I have done the whole thing." We might also, he added, persuade a member of the Royal Family to be a patron. He proposed to invite a prominent Mexican to represent the company in Mexico in return for free shares rather than a salary. Finally, he added that Dr. Mackie "is the foundation of the whole of this scheme."

Dr. Patrick Mackie is an obscure figure about whom very little is known.[19] He had apparently lived in Mexico for several years and was acquainted with many of the leading figures in the country at the time of its independence in 1821. In 1822, he returned to Britain and persuaded the Foreign Office that his experience and local knowledge made him an ideal person to initiate contacts between the British and Mexican governments. Following meetings with the Foreign Secretary, George Canning, he was given a special commission to go to Mexico on what was largely a fact-finding mission. He sailed from Plymouth in December 1822 and after several months in Mexico, he returned again to Britain, arriving back in November 1823. It seems that he had exceeded his instructions and the Foreign Office promptly ended its relationship with him. For his part, Mackie quickly became involved in the rush to form the Mexican mining companies, and using what certainly seem to have been "a cocky determination and persuasive manner," he was soon involved in several projects.[20] Adams seems to have been impressed by him and especially by his offer to use his alleged friendship with Mexican President Guadalupe Victoria to smooth over any difficulties.

Adams made several other promises about appointments to the company and recruitment of the best miners in England. His letter and subsequent ones were addressed to General Wavell, in whose papers they are now to be found.[21] Wavell had probably met Adams in the spring of 1823 when he was in London, seeking investors, as explained earlier, for the mining industry. The two became close allies in both commercial and personal ventures.[22] After some months in London, Wavell returned to Mexico with what seems to have been a commission to acquire silver mines for Adams and his backers. In the letter cited above, Adams urged him to find and get control of mines quickly and told him that he was to be appointed the Engineering Commissioner in Mexico for the new Mexican Mine Company at a salary of £1,500 a year.

Two weeks later, 8 January 1825, Adams wrote again to "My dear General." He confirmed his appointment as Engineering Commissioner and gave him various details of Mexican mines he wanted him to investigate. Within a few weeks, however, there was some disagreement between them—no details are known—and Adams wrote to Wavell telling him that he was dismissed from the company. He was to hand over control of any mines belonging to it to a Mr. Henderson who would deliver the letter in person. This was William Henderson who arrived in Mexico in late 1825 or early 1826 and assumed the post of director of mines. He worked for the company for a short time before taking up a post in one of the other British companies.

Adams also told Wavell of another person now involved in the business. This was William Bullock. As we have seen earlier, he had spent several months in Mexico in 1823, returning to Britain in November of that year. One of his traveling companions on the return voyage was Dr. Mackie. While in Mexico Bullock had acquired the silver mine of Milan or Del Bada at Temascaltepec in the state of Mexico. At this early stage, restrictions on foreign control of mines had not yet been lifted and to get around this difficulty, Bullock, with the help of Lucas Alamán, had been allowed to take out naturalization papers.[23] Once again the details are not clear but it appears that he sold his mining interests to Adams and Mackie and the Mexican Mine

Company that they were establishing. In his letter of 16 December to General Wavell, Adams states more than once that Bullock's mine has been "engaged" and he asks Wavell "to gain really correct information about Bullock's mine." In a subsequent letter of 8 January 1825, Adams wrote: "Bullock retains one third clear profit without any advances of his own and receives £5,000 as a premium for the advances he has made and his personal time and trouble." According to Ward, Bullock and his son were appointed agents for the company in Mexico. His salary was £700 a year and he was provided with enough funds to pay the travel expenses to Mexico of himself and his family, fifteen miners, a smelter, and a gardener. Bullock invested in another nearby mining property that he called St. George; expensive machinery was acquired; new adits were dug and extensive drainage works were implemented. Notwithstanding his efforts, however, and those of his son and William Henderson, who managed the mining operations at Temascaltepec, it was soon clear that no silver was there to be found. Ward visited the mine in 1826 and reported that it was "a worthless speculation on which £30,000 had been spent and not an ounce of silver produced."[24]

The Mexican Mine Company, therefore, was not a success. Adams, who seems to have provided much of the capital, lost his money as did the other investors like Rudolph Ackermann who may have invested a modest £200, and Alexander Baring who invested £5,000 of his bank's money and who, according to Ward, was a partner in the venture.[25] Ward also names Sir John Lubbock as a principal backer and there is some evidence that he was closely involved. In 1825, Bullock was preparing to leave for Mexico to supervise the mining operations—his son William was already there—and in his absence from London, Lubbock looked after his financial affairs. In August 1825, Bullock was in Liverpool and in a letter to him on the 25th, his nephew gave him details of bills that had to be paid. He had taken them to Lubbock but he had declined to provide the money. Referring to "the money for the mine," Lubbock had said, "when they received the legal instruments by which they might see they had possession," Bullock would be entitled to the sum they had agreed "but not till then."[26]

In the meantime, Adams attained some fame with the publication of two pamphlets on Mexican silver mining and the British companies investing in it. He was certainly proud of these, as an anecdote in Lord Colchester's diary reveals.[27] On 2 June 1826, a coach left London with several passengers on board, including the poet, Robert Southey. At Deptford, another passenger, Adams, joined the travelers and proceeded to regale them with tales of his Mexican adventures and investments. He condemned public lotteries "but this enemy to gambling soon announced himself as a speculator in South American mines, and almost as the original founder of one of these projects." Moreover, he claimed to have written a book about the South American mines and he was convinced that large profits would eventually be made from them.

Another mining enterprise with some similarities to the Mexican Mine Company was the Catorce Mining Association. This was also formed in 1825 as a private venture and, in the initial stages, no shares were offered to the public. The original investors seem to have included John Cazenove, of the merchant banking family; John Fuller, probably a merchant but possibly also connected to the brewers, Fuller, Smith and Turner; Henry Luard, a West India merchant; Adolphus Goldsmidt, of the firm B. A. Goldsmidt and Co; John Routh, of the merchants Routh and Thompson; Charles William Stokes, possibly a stockbroker with connections to another prominent merchant concern of the same name and a director of other Mexican mining companies;[28] and, finally, Colonel Daniel Robinson. They acquired shares or interests in mines in Querétaro and elsewhere but their main concern was several mines in the Real del Catorce, or mining district of Catorce, located some 9,000 feet in the mountains to the northwest of the Mexican state of San Luis Potosí. These had been successfully worked by Mexican miners in the eighteenth century but like many similar mining areas, had been mostly abandoned during the war of independence.[29] Other British mining companies also obtained mining rights in the area and there was certainly some cooperation between them. Mining tools, for example, were ordered for the Catorce Company in 1825 from the Cornish foundry of Messrs. Harvey and Hayle by John Rule, who was a commissioner for the British-owned Real del Monte Company.[30] The Real del Monte Company also bought machinery from another rival British-owned company, the Anglo-Mexican (see below).[31]

Although very little information about the company's activities is available, recent research by Professor Monroy in local San Luis Potosí archives has identified a number of the company's British employees who arrived in Catorce during 1825 and 1826.[32] It was also in that year that Henry Ward first inspected the company's operations. He noted some problems and was unable to obtain a progress report from the local management because of "the disorganization of the Catorce Company, occasioned by the failure of the House of Goldsmidt."[33] Here he was referring to the failure in February 1826 of the B. A. Goldsmidt merchant and banking firm. Little is known about this firm and none of its records appear to have survived. It was one of many partnerships operating in London at the time and it had extensive commercial and financial interests in Spain and Spanish America, Holland, Germany, and Russia. With contracts already for Portuguese and Colombian loans, it had also won the contract, against fierce competition, to organize the Mexican government's first loan in January 1824. We do not know the nature or extent of its connections with the Catorce Company but as noted above, Adolphus Goldsmidt was a partner in both companies. At any rate, it seems that the collapse of B. A. Goldsmidt brought serious financial problems to the mining venture. Ward, however, was not pessimistic about the future. He noted that the investment to date was around £30,000 to £40,000 and estimated that another £50,000 was needed. If that could be found, he thought, profits of 60 to 70 percent could be expected for years to come.[34]

Before leaving the Real del Catorce, there is one further example of British involvement. One of the richest mines in the area had been the Concepción mine. Its owner was a wealthy Guanajuato miner, Colonel Ignacio Obregón. He rented or sold the working rights to the mine to the Mexican merchants, Murphy and Gordoa. In 1821, they contracted a British engineer, Robert Phillips, to obtain a large steam-driven pump engine, transport it from England to Mexico, and install it in the Concepción mine. The story of what turned out to be an almost epic journey by Phillips was told by him in his own account published first in 1827 and several times subsequently.[35] In brief, he left London on 11 November 1821 with the machinery. After numerous adventures and overcoming immense obstacles in transporting the machinery in sections loaded onto fourteen carts across mountain ranges in Mexico, he arrived at Catorce exactly one year later, on 11 November 1822. Mechanical and financial problems delayed assembly and installation of the pump but eventually it was started in January 1824, only for it to be discovered that the wooden pipes being used were inadequate. Phillips left for the United States to acquire iron pipes, leaving behind the only other Englishman who had accompanied him, a Mr. Newhale, who apparently was drunk every day. Phillips was back in Catorce with the iron pipes by February 1826 and the steam pump so laboriously transported from England was finally brought into operation. Again, financial problems promptly intervened and work stopped in November 1826.

The Association for Working the Mines of Tlalpujahua and Others in Mexico, to give it its full title, was founded in November or December 1824. We have some details of its original promoters thanks to an anonymous pamphlet, signed "A Merchant," published in 1825 and which quickly ran to three editions in the first year.[36] This tells us that Count Vincent de Rivafinoli traveled to Mexico in 1824 as an employee of the Real del Monte Company (details of this company follow below) but for various reasons returned to Britain after only a few months. While in Mexico he had reached an agreement with several mine owners in the Tlalpujahua district in the state of Michoacán, about two days' journey from Mexico City.[37] This authorized him to set up a company to work the mines that like so many others in Mexico had been abandoned for several years. In London, he contacted the merchant partnership of George Green and William Hartley, who agreed to organize the new company.[38] Directors were found and advice was taken from the Mexican diplomat, Mariano Michelena, whose family estates were near the Tlalpujahua district. With these arrangements in place, a public statement was issued on 22 December 1824. It announced the intention to form a company subject to a preliminary inspection of the mines. 'Competent persons" were to go to Mexico, "to take legal possession of them, and to examine on the spot their nature and actual state." The directors, it added, were confident that there was a fair probability of success but "they were exceedingly anxious to make their Shares not transferable till more accurate information could be obtained than that which they now possess." This statement was being made, they concluded, to avoid the shares acquiring any "fictitious value, owing to the excited state of the public mind, at the present moment, with regard to all Mining Speculations."[39]

The Tlalpujahua Company, therefore, was at first a private venture. For a short time afterward, there was a dispute over the rights to some of the mines. This delayed progress until 20 January when the prospectus was published. The capital of the company was to be £400,000 divided into 1,000 shares of £400 each. The first installment was £20 per share with the remaining £380 called for at the directors' discretion. The directors included George Green and William Hartley; William Sampson, also a merchant; John Cowell, probably a brewer of Camberwell Green; George Warde Norman, a director of the Bank of England for more than half a century from 1821 to 1872 and a prominent figure for many years in London's financial community. His family business was involved in the timber trade with Norway. The bankers were Smith, Payne and Smith, a bank that was heavily involved in the joint stock companies, acting for at least nine of them, including the Río de la Plata Mining Company. Two of the bank's partners were Tlalpujahua directors. These were John Smith, Member of Parliament at the time for Midhurst, and his nephew, George Robert Smith, who later succeeded him in the parliamentary seat. Both were highly respected bankers, and the Smith family was well connected socially with links to the Wilberforce family and those of several aristocrats. They were also, it may be noted, shareholders in some of the other British Latin American mining companies.[40] Finally, on the Board of the Tlalpujahua Company, there was a Colonel Henry Cooke, about whom nothing is known.

Within just a few weeks, the first party of miners left for Tlalpujahua. Unusually, these were not largely drawn from Cornwall. Rivafolini had traveled to Europe on a recruiting mission and he attracted several nationalities. One was Mark Beaufoy, formerly of the Coldstream Guards, and in an account of his experiences that he later published, he tells us that on the ship that left Portsmouth on 12 March 1825 were Belgian, French, German, and Italian recruits.[41] John Charles Beales, originally from Norfolk, was appointed the company surgeon. The chief mining superintendent was the sixty-one year old Count de Waldeck, and a German miner, Joseph Burkart, was the director of operations. He too was to write an account of his time in Mexico.[42] The company's links with Germany were also seen in the purchase of equipment from Holtzappel and Co. A year or so later, in September 1826, in what was in effect a publicity stunt that attracted much attention at the time, the company paraded a group of German miners "dressed in blue smock frocks" in the London Stock Exchange prior to their sailing for Mexico in a group of about forty new employees.[43]

One of the largest of the British Mexican mining companies in terms of its capital and operations was the "Association for assisting in working the mines of Mexico and other parts of Spanish America," a title that was soon abbreviated to the Anglo-Mexican Mining Association or Company. It seems to have been formed in the autumn of 1823 with the first public announcement early in December.[44] Contracts with Mexican mine owners had been signed through an intermediary. This was William Dollar, an Englishman resident in Guanajuato in 1823. He had been commissioned by various local mine owners to seek British investment in their mines and he had traveled to London in July 1823. There he contacted and came to agreements with various merchants such as Barclay and Richardson and Powles and Co. and the mines contracted by them formed the basis of the new company.[45] In return for his efforts, Dollar was given 240 shares and paid expenses amounting to £4,450.[46] The prospectus was promptly available and copies dispatched to enquirers all over Britain. It provided what became the standard information for all the later companies. The Association had been formed, it said, to provide the capital required to work rich but abandoned silver mines in Mexico. The contracted mines were listed, most being in the Guanajuato district, and some in San Luis Potosí. The best known was the Valenciana mine which, investors were told had yielded almost 14,000,000 dollars in the years between 1794 and 1802, which after production expenses had left an annual profit for the proprietors of 643,479 dollars. Furthermore, "It is believed, that by the introduction of English capital, skill, experience and machinery, the expenses of working these Mines may be greatly reduced, and their produce much augmented." Within two years, profits were expected to be available for distribution to shareholders. The capital of the company was £1,000,000 divided into 10,000 shares of £100 each, on which a first installment of £5 was payable. Other clauses concerned voting rights of shareholders, meetings, the appointment of a secretary, and the Deed of Association.

The bankers were the well-established Barclay, Tritton, Bevan, and Company and the auditors were Thomas Richardson, a founder of the bill brokers, Overend and Richardson, and William Fry, a partner in the bank of Frys and Chapman. Both bankers and auditors, it may be noted for future reference, were Quakers. There were twelve directors, all of whom were prominent figures in London's financial and political circles. Three were Members of Parliament—Matthias Attwood, Stewart Marjoribanks, and William Thompson. Attwood was from a rich family of ironmasters and bankers. Marjoribanks was a successful merchant and director of many companies and Thompson, another wealthy ironmaster, was a director of the Bank of England, and Lord Mayor of London in 1828. The first chairman was David Barclay. A well-connected merchant, he was a member of the brewing family of Barclay-Perkins and he was elected to Parliament in 1826. In partnership with another director, Charles Herring, he was to win the lucrative contract to arrange the Mexican government's second loan in 1825. Other merchant and banking companies represented on the Board included James Hughes Anderdon, David Bevan, Richard Mee Raikes, George Lyall, Benjamin Shaw, and William Ward, also elected to parliament in 1826. Finally, there was John Diston Powles, who with David Barclay seems to have been the main promoter of the new company. He was to have a long, and controversial, business life. For our purposes, we need only note that at this early stage in his career, he was active in trade and commerce with the Caribbean region and with other partners, including Charles Herring, had won the contract for the first Colombian loan in 1822.

All of the directors, bankers, and auditors were experienced and respected members of London's financial, mercantile, and political elite. It was, therefore, no surprise that the company was an immediate success. The shares quickly rose to a premium of £15, that is, the £100 share on which £5 had been paid was for sale at £20, and there was no shortage of buyers. Experienced miners were recruited together with all the other diverse skills—doctor, chemist, architect, carpenter, and accountant. Thomas Garby, a Cornish mining captain or manager, was named as the "Director of Underground Works" and retired army colonel Daniel Robinson, and John William Williamson, son of the recorder of Newcastle-upon-Tyne, were appointed as the company's first commissioners in Mexico. Writing to the British Foreign Secretary to ask for diplomatic support and letters of introduction to British diplomats in Mexico, David Barclay emphasized that "special instructions are given to the Commissioners to abstain from all interference in the Politics or religion of the Country."[47] Ships were hired, mining equipment bought, including a beam engine from the Quaker-owned Neath Abbey Ironworks of Glamorgan, and the first group of employees left Falmouth early in April, arriving in the Mexican port of Alvarado on 18 June 1824. Robinson and Williamson immediately wrote to the Mexican government to announce their arrival and to ask for official help and protection of their activities. They also enclosed a list of the names of the twenty-six men who had accompanied them.[48]

The expedition went directly to Guanajuato and at first everything seemed to go well. In addition to those already contracted in Guanajuato, shares were acquired in a large number of other mines in the states of Mexico, Querétaro, and San Luis Potosí. In September 1824, Captain Garby sent back an encouraging report, assuring the directors and shareholders that their investments would be profitable as a result of "the introduction of our improved Cornish system of mining."[49] Press reports in London noted that things "are said to be proceeding favourably," with the construction of buildings, smelters, and other ancillary works going well. Local people in Guanajuato and elsewhere had welcomed the arrival of the British into their communities; "the disposition of the people towards us is excellent."[50] Calls for additional funds from the shareholders caused no difficulties.

The United Mexican Mining Association originated in Paris, where Lucas Alamán had gone in April 1822 after attending the Spanish parliament in Madrid as a delegate of his native province of Guanajuato. Anxious to attract European capital into the Mexican silver mining industry, and believing that France was a good place to look for investors, Alamán quickly organized his first company. This was Vial, Alamán and Co., the main partner being Nicolas de Vial y Eydelin.[51] Little interest seems to have been shown in it and Alamán decided to start another venture. This was called the Franco-Mexican Company, which was to be capitalized at 6,000,000 francs divided into 6,000 shares of 1,000 francs each. Its prospectus made the usual claims regarding Mexico's gold and silver mines that for centuries "have poured their precious metals over the globe" and have been "the origin of immense private fortunes." The vein of Guanajuato alone had yielded £54,000,000 from 1770 to 1810 and yet only one-eighth of its extent had been mined. Humboldt was quoted, large and quick profits were promised, and the public was assured that "Never, perhaps, was an Association for purposes of industry, revenue, and trade, projected on a more certain and more profitable speculation."[52] Notwithstanding such assurances, the French apparently showed little interest and Alamán decided to promote his company in the British market. Hence, he spent about six weeks in London in July and August 1822 and reached agreement with the merchant firm of Hullett Brothers and Co., of 102 Leadenhall Street.

Hullett Bros. agreed to market the shares and seek British investors but realized that the central management and control of the company would have to be in Britain. Alamán had returned to Mexico, where soon after his arrival in March 1823, he was appointed to the senior cabinet post of Minister of Home and Foreign Affairs. Before leaving London, he had authorized a Spanish lawyer, Vicente González Arnao, to act for him. Although we have no details, it was probably this lawyer, in conjunction with Hullett Bros., who conducted the negotiations that led to the formation of the United Mexican Mining Company which in effect took over the Franco-Mexican company. The first mention of it came early in January 1824 when Hullett Bros. placed an announcement in the press that "necessary modifications to the constitution" were causing a few days' delay but that the prospectus would soon be ready.[53] It was, in fact, circulated in February 1824. The new company was to have £240,000 in capital divided into 6,000 shares of £40 each on which the first installment was £5. The twelve directors included some very prominent names all with directorships in a number of the joint stock companies. Several were bankers: John Biddulph, Rowland Stephenson, Samuel Bosanquet, James Heygate, Jr., and John Lavicount Anderdon. Merchants included John Hullett and Frederick Pigou, of Pigou, Andrews and Company, gunpowder merchants. There were also two stockbrokers—Jacob Ricardo and John Easthope, later the owner of *The Morning Chronicle*. It is noticeable that unlike most of the other joint stock companies, there were no serving Members of Parliament. The bankers to the company were Bosanquet, Pitt, Anderdon and Co., a long-established and highly respectable Lombard Street establishment, later absorbed into Lloyds Bank.

The presentation of the company was slightly unusual. Unlike the other companies, the prospectus emphasized that it was to be a joint venture "with Mexican interests." Lucas Alamán was named the principal Mexican partner and head of a Board of Management in Mexico. With the exception of two managers and any other assistants it was thought necessary to send from Europe, all other officers were to be appointed in Mexico by the Mexican Board. No contracts with Mexican mine owners had yet been signed because it was intended to rely before doing so on "the local knowledge, practical experience, extensive research and prudential character of Don Lucas Alamán, who has long been a Mine proprietor." Other clauses in the prospectus cover the usual arrangements for meetings, shareholders' voting powers, distribution of profits with the first dividend promised as soon as a 5 percent profit on the investment had been made.

Over the next few months of 1824 and early 1825, dozens of mines were contracted on behalf of the company by Alamán. He personally inspected some, especially in his home state of Guanajuato, and employed agents to check on others in the states of Mexico, Valladolid (Michoacán), Zacatecas, and Oaxaca. The contractual arrangements varied. In Guanajuato, for example, some of the agreements were as follows:

> San Juan de Rayas—one-fourth for ten years. Contract dated 27 November 1824.
> Sechó; five-eighteenths in full property. Contract dated 18 November 1824.
> Cata; one-sixth to 31 December 1836. Contract dated 29 March 1824.
> Calera; one-fourth in full property. Contract dated 5 March 1825.
> San Juan; eleven twenty-fourths in full property.[54]

Alamán also sent reports to London describing and explaining in detail the technical mining problems and potential of each property. He was always careful to emphasize the probable wealth to be found. Shareholders were told, for example, that "The latest advices from Mr. Alaman [sic.] (10th June) state that at the Calera mine, according to the last post from Guanaxuato [sic.], they were already on the lode, in sinking the pit, and had pierced veins running out of it which were very rich, and that every appearance indicated the best results."[55] Past bonanzas were often mentioned as was the expected high yield of silver per 100 lb of ore. The abundance of local labor and of natural resources needed for the mines, such as water, timber, and grain to feed livestock and people were noted. In short, shareholders were given a comprehensive, if slanted, picture of the mining districts into which their money had been or was being invested. As the directors concluded in their first published report of 7 September 1825, "the returns to Share-holders will be fully commensurate with the expectations they may form, upon the perusal and examination of this Report."[56]

The United Mexican Mining Association, therefore, was offered to the public as a judicious investment promoted by cautious, responsible people who were not known for taking undue risks. The same may be said of the Company of Adventurers in the Mines of Real del Monte, or Real del Monte Company, as it became known. It is perhaps the best known of all the joint stock companies formed to invest in Mexico because of a number of factors. There is, for example, a unique and rich archive of the company's records. This has permitted many scholars to study its activities and as a result, again uniquely for the British mining companies, there is a monograph devoted to its history.[57] The archive is housed at Pachuca, in the state of Jalisco, and both it and the nearby town of Real del Monte retain many buildings, monuments, and other reminders of the British presence, not least an English cemetery and the still popular Cornish pasty brought to Mexico by the Cornish miners.[58] Finally, in the context of the 1820s, the Real del Monte Company has been used by countless historians to illustrate the extremes of the stock market Bubble because its shares at one very brief point rose to more than £1,300.

We also know in some depth the details of the company's origins because of a legal case argued in the Court of Chancery in 1824.[59] According to the evidence given in Court, early in 1823 Thomas Kinder, a London merchant, wrote to his business partner in Mexico City, Robert Ponsonby Staples, to look into the possibility of contracting the mines belonging to the Count of Regla, which were located in and around Real del Monte. Staples replied in August 1823 that he had reached agreement with the Count whereby Kinder was given his full authority to look for British investors. Kinder then contacted John Taylor, a well-known mining engineer with extensive connections in the Cornish mining communities and also among wealthy businessmen in his native Norfolk.[60] Taylor agreed to the formation of a company that would have £200,000 capital divided into 500 shares of £400 each with an initial payment of 10 percent or £40. He was to retain 300 shares for himself and his friends, and Kinder was to have the remaining 200 shares.

These discussions probably took place in the final weeks of 1823 and on 16 January and 4 February 1824, the first meetings of interested parties took place. At these, Taylor was formally authorized to contract the Regla mines and also those at a nearby property, known as the mine of Morán. A principal shareholder in this mine was Thomas Murphy. Over the next few weeks, Taylor concluded agreements with Kinder, representing the Count of Regla, and with John Murphy, acting on behalf of his brother Thomas. The various Regla-owned mines at Real del Monte were transferred to the control of the new company for a period of twenty years—later increased to twenty-one—and subsequent negotiations added clauses concerning such matters as the division of profits, the use of British machinery, and an annual payment to the Count of £2,400. In the arrangement for the Morán mine, the contract was for twenty-one years during which an annual payment of £2,000 was to be made to Thomas Murphy "in advance of the Profits, and also the surplus Profits above that Sum in respect of Five Barras or 5/24 Parts of my interest in the said Mine."[61]

Thus, the Real del Monte Company was established, initially, as a private venture. No prospectus was published, but a Deed of Association was drawn up and signed on 16 August 1824. It contained the usual clauses concerning meetings, voting rights, shares, and shareholders.[62] With contracts and directors in place, arrangements were quickly made by John Taylor to send men and equipment to Mexico. Three commissioners were appointed to manage the mining operations.[63] They were James Vetch, formerly of the Royal Engineers, John Rule, a Cornish miner, and Vicente Rivafinoli, who was apparently a friend of the Count of Regla. Vetch and Rule continued to work for the company for several years but Rivafinoli, as we have seen earlier, soon returned to London, where he was involved in the formation of the Tlalpujahua Company. They led the first expedition to sail for Mexico, leaving Liverpool on 25 March 1824 with twelve companions including mine captains, a blacksmith, a carpenter, and other skilled personnel. On arrival at Real del Monte on 11 June, they conducted a detailed survey of the mines, acquired several others, and also took possession of "a considerable tract of land, comprising four large farms." These properties, comprising some 16,000 acres, were leased to secure supplies of grain and especially timber required for the use of steam engines. In general, it was reported "everything was going on very favourably. The kindness and civility shown by the people of all ranks to our countrymen is stated to be unbounded."[64]

While first steps were being taken in Mexico, John Taylor was busy engaging more employees and ordering equipment. In his report of 28 February 1825, he listed the company's intended establishment as being ninety-five persons. They included those involved directly in mining—metallurgists, surveyors, engineers, underground captains—but also a wide range of other occupations such as medical staff, no less than eight bookkeepers, nine blacksmiths and boilermakers, a hedger, a gardener, and a farmer, one John Hope. A huge range of mining equipment was ordered, with an aggregate weight, according to English, of at least 1,600 tons.[65] There were nine steam engines, five for pumping, two for stamping ore, and two for sawmills; various pumps; iron works; "tools and implements for engineers, millwrights, miners, smelters, carpenters, coopers, blacksmiths, farriers, wheelers, and harness makers"; 150 wagons; and 760 sets of mule harnesses. Three ships were commissioned to transport the equipment and an additional twenty-eight people were hired to take it from the port of arrival in Mexico to the mining area.

The Real del Monte Company, therefore, commenced its operations in 1824. Managed by experienced and well-qualified mining experts and what was said in *The Times* (6 October 1826) to be "sober, discreet and worthy persons," it was widely expected to be a successful venture. The same opinion was expressed about the Bolaños Mining Company. In most respects, a companion company of the Real del Monte with much the same directors and management, it nevertheless remained a separate organization for more than twenty years. Its origins were in the arrangements which Kinder and Taylor had made regarding the Real del Monte Company and it was their disagreement over what had been resolved that was taken to court for settlement. In brief, according to Taylor, in April 1823, he had been contacted by William Dollar, whom we have already encountered acting on behalf of Mexican mine owners. Dollar claimed to have the owner's authority to contract mines in the Bolaños district to a British company. Taylor accepted the offer and set up the company. Captain Vetch, the Real del Monte commissioner in Mexico, was asked to have the contracts ratified by the mine owners. After some problems, this was eventually achieved with mines leased at Bolaños and also at Zacatecas. Consequently, in January 1825, The Company of Adventurers in the Mines of Bolaños was formally established and a prospectus published.[66] The capital was to be £200,000, divided into 500 shares of £400 each with a deposit of £20 per share. Much of the rest of the prospectus was concerned with the share transactions and some unusual clauses were inserted. For example, no shareholder was to hold more than twenty-five shares "unless he may have acquired them by marriage, will, or in course of administration." Shares were not transferable without the consent of the directors and could not be transferred after a call until the call had been paid.

The personnel were much the same as the Real del Monte Company. John Taylor was to be the manager of both companies; Martineau and Malton were solicitors for both; and the directors were, with one exception, the same. The exception was Sir Robert John Harvey, a partner in a Norwich bank. The Bolaños Company bankers were not the same. Rather than Frys and Chapman, bankers to Real del Monte Company, it was preferred to use Hoare, Barnetts, Hoare and Co., 62 Lombard Street.[67] Like Frys and Chapman, however, the Hoare and Co. partners were closely related to several of the Real del Monte directors—Samuel Hoare, for example, was Thomas Buxton's brother-in-law. Both banks, it may also be noted, were owned by prominent Quaker families.

The obvious similarities between the Real del Monte and the Bolaños Companies were also to be seen in the share allocation. The Real del Monte shareholders, largely the directors, were offered the right to buy the Bolaños shares on a one for one basis, that is, for each Real del Monte share, they were entitled to a Bolaños share. Hence, the principal shareholders in both companies were the same. The dispute with Kinder was over the allocation of shares. He claimed in court that Taylor had refused to give him Bolaños shares on the one to one basis, thereby depriving him of fifty shares. Perhaps in normal circumstances, such a small number of shares may not have merited an expensive legal case but at the time of the dispute—February–March 1825—the Bolaños shares were fetching a premium of £400 to £500. Substantial sums were at stake. Kinder, however, lost his case. Years later in the mid-1830s, he recalled the dispute with Taylor in a letter to Lord Brougham. He claimed to have been the founder of the Real del Monte Company and that he had invested £25,000 in it. He had kept his shares, he said, but because of the dispute, he had not participated in any way in the management of the business.[68]

The Bolaños Company, therefore, was established in February 1825 and although it remained independent of its Real del Monte counterpart, with separate meetings and published reports, both ventures retained close links for many years and for the most part, shared the same directors. The founding directors of the last of the mining companies also stayed in office for many years. This was the Mexican Company. It was announced in February 1825, at the height of the market Bubble, but perhaps because it was in competition with already established companies, its prospectus contained a number of novel points. First, it was not simply a mining company. The intention was to engage in "operations of a general nature" in Mexico and adjacent countries, including mining, trade, and for the purposes of commerce, the cultivation of products indigenous to the region. Hence, in the early months, the company was referred to as the Mexican Trading Company or Mexican Mining and Trading Company.[69] Secondly, prospective investors were told that although Mexican mineowners had already offered their mines, no contracts had been signed. Instead, commissioners "of integrity and knowledge" were to be appointed. They would travel to Mexico, evaluate all offers of mines, and refer them back to the directors in London for decision. In this way "improvident contracts will be avoided." One thousand shares were to be reserved for distribution among mine owners, planters, merchants, and other residents in Mexico who could pay for them with "Bullion, Indigo, Cochineal, or other produce."[70]

The capital of the company was to be £1,000,000 divided into 10,000 shares of £100 each, with a first installment of £10. We do not know who the main promoters were but the intended mercantile as well as mining interests probably influenced the composition of the Board of Directors, most of whom were London merchants. The chairman for several years was John Mitchell. Described by a shareholder as "a man of property," little is known about him but it is likely that he was the senior partner in the merchants, John Mitchell and Co. at 52 Old Broad Street. His son, listed as John Jr., was also a director. Other merchants included Ambrose Obicini, John William Buckle, Edward Burmester, Aaron Chapman, John Exter, Richard Hart Logan, Philip Monoux Lucas, John Petty Muspratt, and Henry Usborne. Unusually for the joint stock companies, there were no politicians or Members of Parliament and the only titled director, Sir Stephen Sharpe, died soon after the company was founded. Similarly, the banking community was not represented, except in the person of Henry Milnes Thornton, who may have been related to the Pole, Thornton and Co. Bank, but he resigned within a few months, possibly because the bank was among those which failed in the crash at the end of 1825. The only notable exception to the merchants was James Deacon Hume, described in the *DNB* as a civil servant and free trader. An expert in customs law, he had achieved fame and monetary reward by reducing some 1,500 acts dating back to Edward I to ten bills that were given the royal consent in 1825. Parliament had voted him £6,000 in recognition of this work and at least some of that money was invested in the Mexican Company. The company bankers were Masterman, Peters, Mildred, Masterman, and Co. Like all the other banks associated with the Mexican companies, it was a long- and well-established business, largely owned by the Mastermans, a highly respected Quaker family.

Shares in the new venture were offered for sale early in March 1825 at a premium of £16 on the £10 paid deposit price. The premium was not surprising as demand was very strong with applications received for 34,000 shares. Exactly how many of the original 10,000 shares were eventually released in the market is not known and soon became a matter of bitter dispute between the directors and disappointed shareholders. For the first few months, however, all went well. £86,950 was raised from the initial share sale—8,695 shares sold at £10 deposit. A London office was opened in Great Winchester Street and two commissioners were appointed, at generous salaries of £2,000 a year each, and they left for Mexico to look for mines and other business opportunities. These were Richard Exter, brother of the director, John Exter, and John Martyr. They quickly came to agreements with mine owners in several places but especially in Oaxaca where, after negative evaluations of mines in other areas, it was decided to concentrate the company's efforts and money. One exception was in Veracruz, where copper and gold mines in the district of Somelahuacan were contracted. Merchandize, machinery, quicksilver, and general mining supplies were bought and shipped at a cost of over £36,000, and within a year, £1,148 had been spent on "rent, taxes, servants, stationery, postage, advertisements, printing, translating documents, coals and minor disbursements."[71] Further calls on shareholders, one of £5 and another of £3 per share were made, bringing in £56,144, and by the end of 1826 the balance sheet showed income and expenditure of £143,094.

Two other features of the Mexican Company may be mentioned at this point. First, the original idea of being a trading as well as mining business was abandoned. It was soon discovered that opportunities for trade were not as readily available as had been anticipated. The shareholders were told that the directors, who met more than fifty times in the first year, had concluded that "trading in Mexico was attended with great risk, expense and expenditure of capital, with long and uncertain returns."[72] Hence, it was decided to abandon the idea of trading in commodities and other items and to concentrate entirely on mining. Secondly, unlike most of the other companies that recruited their managers and miners in Cornwall and other British mining communities, the Mexican Company found its workers in Germany.

Details are scarce but it seems that the first miners sent to Mexico encountered difficulties. Mines that had been leased were too expensive to work or had insurmountable technical problems. It was decided to abandon them and the directors sought advice in London from mining experts. One of these was Aristides F. Mornay.[73] He traveled to Mexico to evaluate the company's properties. On his return, he presented a report in which he condemned both British and Mexican mining methods in use in Mexico and strongly recommended the employment of German miners and the application of "European modes of working Silver Mines." The directors accepted his recommendations and Mornay went to Germany to recruit German miners for the company.[74] Justus Ludewig von Uslar was appointed as Chief Director of Mines. The commissioner, John Martyr, died within a few months of arrival in Mexico and he was replaced by N. H. Schartau. Alexander Popp was Chief Director of Reduction Works and in total some sixty Germans—engineers, assayers, medical staff, smiths, carpenters, and diverse other trades—were recruited. What the directors described as "a strong and well organized corps of German Miners and Artificers" sailed for Veracruz and more were to follow.[75]

We have now seen, briefly, the array of joint stock companies established in London in 1824 and 1825 for the purposes of developing, and profiting from Mexico's natural resources. Ranging from relatively small-scale operations in whale or pearl fishing to major investments in the silver mining industry costing millions of pounds, the British investor was offered a wide choice of options for his or her money. The foreign government bonds on offer, especially those from the new republics of Spanish America, appeared to be a secure investment offering a good return, and thousands of investors chose that option. Innumerable others, however, for whatever reason, decided to put their money into the joint stock companies. Even then, it was not a simple choice because in addition to the Mexican companies, there were, of course, hundreds of other joint stock ventures promising rich returns from investments elsewhere in Latin America, and the world at large. It is also the case, however, that there was no shortage of voices in the press and elsewhere warning of the dangers of speculating in the stock market. The dramatic stories of vast profits or losses being made in the market attracted the attention of the press and the satirical cartoonists of the day. The London and provincial papers warned repeatedly of the dangers of speculation. A pamphlet published in Edinburgh, in the form of letters to a country cousin, cautioned against the "illegitimate offspring of the wanton demon of speculation" and in equally colorful language, warned that "as a gadfly teases a thin-skinned horse, in the dog days, so do these ever buzzing schemes keep snapping at the citizens of Edinburgh."[76]

The company promoters were quick to react.[77] The first to respond was John Diston Powles, a director of the Anglo-Mexican Company but also involved in several other ventures. In a pamphlet entitled *A Letter to Alexander Baring*, he strongly rejected criticism that Baring had made of the mining companies. Citing Humboldt's published work and also a letter from him as recently as February 1824, together with an array of statistics on silver production in Mexico, he reiterated the view that with British capital and expertise, there could be no doubt that that the Mexican mining industry would be made profitable again. The new Spanish American nations offered the British entrepreneur unlimited opportunities: "the earth and the sea are open to us in that immense region, and will, with due care, prudence and diligence, yield us an abundant harvest."[78]

Powles's case for investment in Mexico and the other new nations was quickly supported by William Adams/Rawson. As we have seen earlier, he was also personally concerned with promoting mining and other companies in Mexico and as he admitted, he and his friends and relatives held large numbers of shares. In two publications in 1825 in which he cited Humboldt, Hall, Alamán, Wavell, Robinson, and others, and using statistics of mining output and the production of the Mexican Mint, he sought to prove that companies like the Anglo-Mexican, United Mexican, and Real del Monte, all of which he referred to by name, were making good progress.[79] There was no doubt, he wrote, that "the most lucrative returns are ensured to the capitalist." The joint stock companies also were "the main spring to effect the early regeneration and prosperity of the New World."[80] He confidently expected British commerce to derive great benefits from Spanish America once the beneficial effects of British capital had had time to work. As for Mexico, "the generous and magnanimous character of the Creole population" ensured that their country was destined to be "a great and rising nation."[81]

Disraeli was another energetic defender of the joint stock companies in 1825. He also was an active, albeit unsuccessful, speculator in mining company shares and in collusion with his friend and mentor, Powles, he composed long pamphlets in defense of the companies.[82] He too rejected the criticism of Baring and Hobhouse and he refuted the charge that the companies were "devised by fraud and supported through delusion, conceived by cheats and patronized by gamblers."[83] On the contrary, he argued that they were sound investments and that the Mexican companies, although only recently formed, were already yielding profits. Again, supporting his points with ample statistics of silver production and citing Humboldt and others, he made the case for investment in the companies, assuring potential shareholders that the superior mining technology of the British, especially Cornish miners, was certain to bring rapid results. In his private correspondence, Disraeli was even more emphatic. In a letter to Robert Messer in April 1825, he assured him that:

> On the Mexican mines I rest my sheet anchor. When I tell you that one single mine which everybody in this kingdom believes to be inundated . . . is producing at the rate of £3000 *net profit* per week and that at the end of the first year of working it . . . it will give us a net profit at the least of £150,000, when I tell you moreover that this mine contains veins of immense and incalculable worth, which have only been discovered by our own surveyors and that it is their opinion that it alone may yield cent (100) pr cent for our money . . . our agents are writing home to us to buy them additional (shares), no matter at what price. Can you call me too sanguine?[84]

It was also in 1825 that the stockbroker, Henry English, published the first of his detailed surveys of all the new joint stock companies. We have referred to his figures in an earlier chapter and all we need to note at this point is the image of Mexico that he offered to his readers. "Mexico," he said, "presents a field of enterprise to the capitalist, perhaps unequaled in any part of the globe, from the vast riches contained in its mountains and the fertility of the country, affording peculiar advantages and facilities."[85]

These writers—Powles, Adams, Disraeli, English—were by no means the only ones to promote the Mexican companies and the image of Mexico as a safe place to invest. John Taylor, for example, published another pamphlet in which he again sought to demonstrate the probable profitability of the Real del Monte Company's Mexican mines.[86] Also, although no new books by British travelers appeared in 1825, there was a well-received and up-to-date description of Mexico by Joel R. Poinsett, who had become the first United States Minister to the new nation. His book, *Notes on Mexico made in 1822*, originally published in Philadelphia in 1824, was promoted by his London publisher in 1825, John Miller of New Bridge Street, as the best available account of Mexico's present situation.[87] Although Poinsett did not especially emphasize commercial opportunities, he included an impressive array of statistics on silver production, largely taken from Humboldt, together with information on the mines and mining districts, concluding, "There can be no doubt, that the product of the mines of Mexico, is capable of being considerably augmented."[88] Perhaps reflecting the growing interest in Mexico for visitors other than merchants and miners, a two-volume work on Mexico was also included in a multi-volume series entitled *The Modern Traveller*, edited by J. Conder. It consisted largely of extracts from the books already published by Adams, Bullock, Hall, Robinson, Poinsett, and the ever present Humboldt. Priced at only 2s. 6d., it was clearly hoped to attract a wider readership.[89] A new map of Mexico, said to be an improvement on Humboldt's, was offered for sale in February 1825, advertised as "Arrowsmith's Map of Mexico in 4 sheets at £1 11s.6d." For those who wanted to know the location of the Mexican mines in which they were investing their money, there was a new map of "Mexico and Guatemala shewing [sic.] the position of the mines" by the royal geographer and cartographer, James Wyld.[90] Finally, a number of official Mexican documents were translated and offered for sale in London in 1825. These included congressional and Treasury reports as well as current legislation on the mining industry.

The press in general also continued to give Mexico favorable comment. Events such as the opening of the first constitutional congress in January 1825 were reported and *The Times* (21, 27 May 1825) described a nation enjoying political peace and growing economic prosperity. President Victoria's speech to congress was quoted: "the army has been paid and the civil list has been satisfied" (18 July 1825). *The Morning Chronicle* (1 April 1825) praised Powles' pamphlet and reported rumors that the Valenciana mine was already yielding profits of 12,000 dollars a week (15 April 1825).

When the market bubble burst in the spring of 1825, the weight of all this pro-Mexico publicity was enough to protect the Mexican companies. While some of the fraudulent ones disappeared, all of the main companies formed to invest in Mexico escaped the consequences of the market panic. Most notably, the mining companies survived. Investor confidence in Mexico's natural resources, above all in its silver mining industry, was dented but only marginally and while the share prices promptly fell from the levels seen at the height of the Bubble, enough capital—almost £4,000,000—was provided by shareholders to enable large-scale operations to begin in mining areas throughout Mexico. All concerned—directors, managers, miners, and shareholders—confidently expected to receive rich dividends on their investments and to receive them very quickly. What happened to the companies and their shareholders? Did they profit from the investment in Mexico? Did the use of British technical expertise applied by Cornish and other experienced British miners and engineers bring about the expected benefits? In short, what was the fate of the first wave of British mining companies in Mexico? It is these and related topics that we consider in the next chapter.

NOTES

1. The quotation is from fol. 184 of vol. 63 of the journals of the Rev. William Skinner. The journal for 1826 from which the quotation is taken is in the BL at Add. 33695.

2. R. W. Randall, *Real del Monte: A British Mining Venture in Mexico* (Austin: University of Texas Press, 1972), 30–31.

3. Letters in SRE, 1-1-44.

4. M. E. Rankine, "The Mexican Mining Industry in the Nineteenth Century with Special Reference to Guanajuato," *Bulletin of Latin American Research* 11, no. 1 (1992): 1:30.

5. See "Statement of contracts for mines entered into by the commissioners of the Anglo-Mexican Mining Association" in the appendix to *Report of the Directors of the Anglo-Mexican Mining Association . . . 3 January 1827* (London: Richard Clay, 1827).

6. See, for example, the advertisement in *The Age*, December 13, 1829, offering one *barra* in the Morán mine.

7. S. Schwartz, "Creating the Cult of 'Cousin Jack': Cornish Miners in Latin America, 1812–1848 and the Development of an International Mining Labour Market." www.projects.exeter.ac.uk/cornishlatin/Creating%20the%20Cult%20of%20Cousin%20Jack.pdf.

8. Details from NA, F.O. 50/43, fols. 7–8. In December 1826, Lees had married into the Otero family, a wealthy Guanajuato mining dynasty.

9. R. Stephenson to William James, April 18, 1824, Liverpool Record Office, Ref. 385, JAM/1/7/1.

10. W. D. Skeat, *George Stephenson: The Engineer and His Letters* (London: Institute of Mechanical Engineers, 1973), 64–69. Correspondence in Institute of Mechanical Engineers, IMS 135/2.

11. See the Harvey and Co. letter book for 1825 in Cornwall Record Office. Ref. H. DD H/2/4. The company built at least three steam engines for use in Mexican mines.

12. Schwartz, "Creating the cult of 'Cousin Jack.'"

13. The list of cargo is in SRE, 2-3-2251.

14. See, for example, *Times*, November 25, 1824.

15. NA, F.O. 50/21, fol. 34. James Nisbett Colquhoun was one of the Real del Monte company commissioners.

16. J. Vetch to J. Taylor, May 28, 1825, Archivo Histórico y Museo de Minería, Pachuca, Mexico (henceforth AHMPachuca), Correspondencia, Exp. 1 (8), fols. 35–36.

17. *Morning Chronicle*, July 5, 1825.

18. Foreign Stock Market Committee Minute Books, Guildhall Library, Ms. 14617/1. Year 1825, fols. 262–63. These letters have been attached to the Minute Book.

19. Details of Dr. Mackie given here are from H. McKenzie Johnston, *Missions to Mexico: A Tale of British Diplomacy in the 1820s* (London: British Academic Press, 1992), passim.

20. Ibid., 3.

21. NA, E192/5.

22. There are some interesting details about Wavell in McKenzie Johnston, *Missions to Mexico*, 37–43.

23. Mexican congressional approval of Bullock's naturalization papers was given on June 26, 1823. The legal restrictions on foreigners in the mining industry were suspended a few months later. See J. A. Mateos, ed., *Historia parlementaria de los congresos mexicanos* (Mexico: Imprenta de J.F. Jens, 1878), 2:452, 455, 531.

24. H. G. Ward to G. Canning, Mexico, September 15, 1826, NA. F.O. 50/23, fols. 160–89. H. G. Ward, *Mexico in 1827*. 2 vols. (London: Henry Colburn, 1828; 2nd expanded ed., 2 vols. London: Henry Colburn, 1829), 2:80. Bullock sent Ward a report on the mining operations in July 1826. At that time he was still optimistic of success but by October 1826 he had given up hope and was preparing to leave Mexico. Bullock's report is in NA, F.O. 50/24, fols. 157–61.

25. It seems to have been generally known that Bullock sold his mining interests to Lubbock and others: see the letter from Hullett Bros. to L. Alamán, August 6, 1825 in Alamán-Hullett correspondence, AGN, Hospital de Jesús, vol. 827. Ackermann invested £200 in the Mexican

Mine Company in October 1824, according to his bank records. Although this could refer tc another company, it is also possible that it was the Bullock concern—Ackermann and Bullock were certainly acquainted. Information kindly provided by John Ford.

26. G. Lackington to W. Bullock, August 25, 1825, in Lackington Letter Book, Hereford Record Office, F58/1. Lackington was Bullock's nephew by marriage.

27. C. Colchester, ed., *The Diary and Correspondence of Charles Abbot, Lord Colchester, Speaker of the House of Commons, 1802–1817* (London: John Murray, 1861), 3:443–44.

28. There are several merchants listed in the London directories with the name Stokes, for example, Benjamin Stokes and Co., Henry Stokes and Sons. There are also stockbrokers with the same name. It was not unusual for such partnerships to act both as merchants and brokers.

29. R. Montejano y Aquinaga, *El Real de Minas de la Purísima Concepción de los Catorce* (San Luis Potosí, Mexico: Academia de Historia Potosina, 1975).

30. Messrs. Harvey and Hayle, Letter Book, Harvey and Co. to John Rule, December 13, 1825, Cornwall Record Office, Ref. H: DD H/2/4.

31. See "Receipts and disbursements of the Anglo-Mexican Mining Association," in *Report of the Directors of the Anglo-Mexican Mining Association . . . August 28, 1826* (London: Henry Stokes, 1826).

32. I. Monroy Castillo, *Sueños, tentativas y posibilidades. Extranjeros en San Luis Potosí, 1821–1845* (San Luis Potosí, UK: El Colegio de San Luis Potosí, 2004), chapters 7 and 8.

33. Ward, *Mexico in 1827*, 2:107.

34. Ibid.

35. Phillips's account was first printed as an appendix to Ward, *Mexico in 1827*, 1:528–47. It is dated Real de Catorce, January 4, 1827. See also J. F. Pedraza, "Robert Phillips. Detalles de un viaje desde Altamira a Catorce," *Archivos de Historia Potosina*, 5, no. 2 (October–December 1973): 104–40.

36. *An Inquiry into the Plans, Progress and Policy of the American Mining Companies* (London: John Murray, 1825).

37. Tlalpujahua is sometimes spelled with an x as Tlalpuxahua. Ward, *Mexico in 1827*, 2:65, says that a total of ninety-three mines were leased; eighty-six in Tlalpujahua and eight in the state of Mexico.

38. The firm of Green and Hartley was caught in the financial crisis of 1826 and suspended payments for a time. It resumed business later.

39. H. English, *A General Guide to the Companies Formed for Working Foreign Mines* (London: Boosey and Sons, 1825), 61–63.

40. The Smith, Payne, and Smith Bank became part of the National Westminster Bank that is now part of the Royal Bank of Scotland. A small archive of the Smith Bank survives.

41. M. Beaufoy, *Mexican Illustrations, Founded upon Facts with Observations upon the Government and Resources of the Republic of Mexico, as They Appeared during Part of the Years 1825, 1826, and 1827* (London: Carpenter and Son, 1828), 1.

42. J. Burkart, *Aufenthalt und Reisen in Mexico in den Jahren 1825 bis 1834* (Stuttgart, Germany: E. Schweizerbart'sche Verlagshandlung, 1826). See also, J. Burkart, "Aufenthalt und Reisen in Mexico (Residence and Journeys in Mexico between the Years 1825 and 1834, etc.). Stuttgart, 1836. Communicated by Major Chartres," *Journal of the Royal Geographical Society* 10 (1841): 544–51. Burkart worked for the Tlalpujahua Company until 1828 and thereafter for the Bolaños Mining Company. He spent nine years in Mexico and traveled widely throughout the mining districts. Waldeck was born in Prague but had become a naturalized citizen of France: see M. R. D. Smith, *Recollections of Two Distinguished Persons: La Marquise de Boissy and the Count of Waldeck* (Philadelphia: J. B. Lippincott and Co., 1878), 60–62.

43. *Morning Chronicle*, September 22, 1826.

44. See the *Times*, December 10, 1823.

45. Rankine, "The Mexican Silver Mining Industry in the Nineteenth Century," 24–25.

46. *Report of the Directors of the Anglo-Mexican Mining Association, August 28, 1826* (London: Henry Stokes, 1826), "Receipts and Disbursements, 1 January 1824—31 July 1826."

47. D. Barclay to G. Canning, April 1, 1824. NA, F.O. 50/8, fol. 125.

48. Letter in SRE, 2-3-2251, fol. 5.

49. Cited in B. Disraeli, *An Inquiry into the Plans, Progress, and Policy of the American Mining Companies*, 2nd ed. (London: John Murray, 1825), 20–21.

50. *Times*, November 11, 1824. The Anglo-Mexican Mint Company, closely associated with the mining company of the same name, was also established in 1825. Details of its activities are given in chapter 7.

51. Details from J. C. Valadés, *Alamán. Estadista e Historiador* (Mexico: Universidad Nacional Autónoma de México, 1977), 137–40.

52. The Franco-Mexican prospectus is in English, *A General Guide*, 30–34. There is a copy of the original French version in BL, 8244.bb.38 (13).

53. A copy of the notice from Hullett Bros. also appeared in the provincial press. See, for example, *Royal Cornwall Gazette*, January 10, 1824.

54. Details from *United Mexican Mining Association. Report of the Court of Directors, addressed to the Share-holders* (7 September 1825) (London: Philanthropic Society, 1825), 28–29.

55. *United Mexican Mining Association. Report* (7 September1825), 39.

56. Ibid., 55.

57. Randall, *Real del Monte*.

58. The archive is the Archivo Histórico y Museo de Minería at Pachuca.

59. Kinder v. Taylor, March 1825, in *The Law Journal* 3 (1825): 68–84. There is a slightly different version of the origins of the company in Randall, *Real del Monte*, 31–42.

60. R. Burt, *John Taylor: Mining Entrepreneur and Engineer, 1779–1863* (Buxton, UK: Moorland, 1977).

61. In 1828, Murphy sold half his annual payment for £1,000 to a Bordeaux merchant. Brian Mills kindly supplied me with a copy of the original certificate for this transaction.

62. English, *A General Guide*, 54–56, gives the text of a prospectus but adds in a note that it "did not originate with the present Company." Disraeli, *An Inquiry*, 24–25, says that as far as he knew, no prospectus was printed. No copy of the Deed of Association has been found but some of its clauses were included in the report of the Kinder court case.

63. These details are from Randall, *Real del Monte*, 35–42 and from J. Taylor, *Report made to the Annual General Court of the Company of Adventurers on the Mines of Real del Monte, held on the 28th of February, 1825* (London, 1825).

64. Taylor, *Report* (1825), 8.

65. English, *A General Guide*, 97.

66. The formation of the company was mentioned in the *Morning Chronicle*, January 25, 1825.

67. The Hoare and Co. bank was became part of Lloyds Bank in 1884.

68. University College London Library, Brougham Ms. 30,656, entitled "Observations on Mexico," undated but probably written by Kinder in 1835.

69. *Morning Chronicle*, February 25, 26, and 28 and March 3, 1825.

70. There is a copy of the prospectus in English, *A General Guide*, 44–46. Original copies are in BL, 8223.e.10 (95, 96).

71. The financial details are from *Report of the Directors of the Mexican Company presented at the second annual general meeting on Thursday, 3rd May, 1827* (London: J. Plummer, 1827).

72. Report of Mexico Company meeting in the *Times*, May 5, 1826.

73. Mornay was described in a court case reported in the *Times*, June 4, 1830, as "a gentleman of large fortune . . . a merchant trading to Mexico or in some way connected to that country."

74. Mornay's report is included in *Report of the Directors of the Mexican Company*, May 3, 1827.

75. For interesting details about the Germans and Von Uslar in particular, see R. J. Willis, *Justus Ludewig von Uslar, and the First Book on Alleopathy* (Dordrecht, The Netherlands: Springer, 2004), 20–29.

76. Anthony Romney (pseud?), *Three Letters on the Speculative Schemes of the Present Times, and the Projected Banks. Addressed to His Cousin Richard, in the Country* (Edinburgh: Bell and Bradfute, 1825), 3, 36.

77. There was no new legislation on joint-stock companies for several more years.

78. J. D. Powles, *A letter to Alexander Baring, Esq. M.P. on the Subject of Some Observations Reported to Have Been Made by Him in the House of Commons, on the 16th March, 1825 in Relation to the Foreign Mining Associations* (London: Cowie and Co., 1825).

79. W. Adams/Rawson, *The Actual State of the Mexican Mines, and the Expectations of the Shareholders of the Anglo-Mexican Mine Association, Being the Substance of a Letter to the Directors of that Company* (London: Sherwood, Jones, and Co., 1825); *The Present Operations and Future Prospects of the Mexican Mine Associations Analysed by the Evidence of Official Documents English and Mexican, and the National Advantages Expected From Joint-Stock Companies Considered in a Letter to the Right Hon. G. Canning* (London: J. Hatchard and Son, 1825).

80. Adams, *Present Operations*, 7, 75.

81. Adams, *Actual State*, 46.

82. Disraeli, *An Inquiry*; *Lawyers and Legislators: Or, Notes on the American Mining Companies* (London: John Murray, 1825); *The Present State of Mexico: As Detailed in a Report Presented to the General Congress by the Secretary of State for the Home Department and Foreign Affairs at the Opening of the Session in 1825. With Notes and a Memoir by Don Lucas Alamán* (London: John Murray, 1825).

83. Disraeli, *An Inquiry*, 4.

84. J. Gunn et al., eds., *Benjamin Disraeli Letters: 1815–1834* (Toronto: University of Toronto Press, 1982), 1:29. Apparently, this letter was only a draft and never sent.

85. English, *A General Guide*, 90.

86. J. Taylor, *Statements Respecting the Profits of Mining in England, Considered in Relation to the Prospects of Mining in Mexico; in a Letter to Thomas Fowell Buxton, M.P.* (London: Longman and Co., 1825).

87. Advertisement in the *Times*, April 30, 1825.

88. J. R. Poinsett, *Notes on Mexico* (Reprint, New York: Frederick A. Praeger, 1969), 176.

89. J. Conder, *The Modern Traveller: A Popular Description, Geographical, Historical, and Topographical of the Various Countries of the Globe. Mexico and Guatemala*. 2 vols. (London: James Duncan, 1825).

90. There is a copy of Wyld's map, dated London, 1825, in the Royal Geographical Society Library, MR. Central America G.19. There are later editions.

Chapter Five

Fate of the Mining Companies: Phase 1

Some evil magician has turned all the silver of Mexico into dross.[1]

When Henry Ward inspected the Catorce Company's operations in 1826, he concluded that if another £50,000 could be found to continue the development of the mines, profits of 60 to 70 percent could be expected for years to come.[2] Not long after his visit, two of the Company's directors were in Mexico to make their own assessment of future prospects. They were Colonel Daniel Robinson and Charles William Stokes.[3] Robinson had previously been an officer in the Spanish army "in the service of the King of Spain" and he had first gone to Mexico in 1824 as a commissioner for one of the other British companies.[4] With obvious local knowledge and opportunity, he may well have been the person who had contracted mines on behalf of the Catorce Company. After a few months inspecting and evaluating the Company's properties, in August 1827 he requested passports from the Mexican authorities for himself, his wife and two children, to return to Britain.[5] Although lack of records means we have to make assumptions at this point, it appears that back in London, Robinson reported on the company's prospects to his fellow investors and they decided to seek to raise further funds by selling shares. Hence, early in 1828 a prospectus was published. This gave the names of the original directors as we have listed previously; the bankers—Lubbock, Foster, Clarke, and Co.; the solicitors—Oliverson, Denby, and Lavie—and details of the mines the company was working. It also offered the usual promises of "very profitable results," claiming that the Catorce district had produced millions of dollars in recent years. £60,000 had already been spent on equipment, clearing the mines, preparing roads, stores, and houses for management and other personnel. To make the whole operation productive and profitable, a further £50,000 or £60,000 was required. It was proposed to raise this by the sale of 6,000 "certificates" of £10 each, payable in three installments starting with one of £5, followed by two of £2 10s. Future calls, potential investors were assured, would not exceed £15.

The Catorce Company's attempt to sell its shares was unsuccessful. There was an initial small premium on the £5 deposit with the price quoted in October 1828 from £5 5s. to £5 15s. but there were few takers. Unusually, at least one certificate has survived. This is for five shares numbered 2681–2685 on which the unnamed buyer had paid the required deposit of £25. The certificate is signed by John Cazenove, Adolphus Goldsmidt, and the company secretary, John Noble. It is noticeable that future calls were to be published not just in the London newspapers but also in the *Journal de Commerce* of Paris and the *Borsen Hallen Liste* of Hamburg.

Notwithstanding the attempts to market the company in continental Europe, the required capital was not forthcoming, and with the original partners unwilling to risk any more of their money because "it does not suit" them, the company ceased to operate, probably in 1829.[6] Presumably, its mines were sold or restored to their original Mexican owners.[7] Cazenove and his partners lost their £60,000 investment.

The Catorce Company was the first of the established British mining companies to fail but it was soon not alone. At first everything had seemed to be going well with the Tlalpujahua Company. Ward had inspected its operations in May 1826 and reported to George Canning that good progress was being made.[8] The local manager had assured him that by early 1827, he expected to be making a profit. On 6 February 1827, a shareholders' meeting was held at the company's office in Broad Street. The balance sheet revealed that £145,186 had been subscribed by the shareholders. Bills in Mexico had absorbed £98,886 and in New York, £11,949; and with incidentals such as outfits for employees in Mexico, legal costs, and other miscellaneous items, there was a balance at the bank of £5,682. Recent mining operations had been much better than anticipated and profits could soon be expected. Over 1,000 men were employed, seventeen separate mines had been opened, new adits sunk and a steam engine installed and working. Distinguished visitors to the mines including the U.S. Minister to Mexico (probably Joel R. Poinsett) had been so impressed that they had bought shares. There were still problems with flooding and illness among the men but the directors had no doubt that their business would be "one of the most fortunate in the annals of mining speculations."[9]

Information of what happened to the company over the next eighteen months is scant. In additional details provided for the second edition of his book in 1829, Ward noted that the directors had sent a Mr. Cameron to Mexico to check on progress. He arrived in Tlalpujahua in December 1827, "with a further credit of fifty thousand pounds" which he was authorized to spend if he felt there was a reasonable prospect of success.[10] He returned to

London by July 1828 and his report indicated differences of opinion over the mining methods being used but common agreement that more capital was needed. At least two calls had recently been made on the shareholders, for £30 a share in January and another £20 a share in April 1828. For reasons that are not clear, the value of the shares collapsed dramatically, if only temporarily, later in April. *The Times* (22 April 1828) reported "one of the most extraordinary instances ever known of the depreciation of property" had occurred when the Tlalpujahua company shares on which £270 had been paid were being offered for sale at 20 to 23 shillings. The price certainly recovered to around £30 but shareholders were still facing heavy losses and according to Ward, in August 1828 a meeting of shareholders, faced with advice from the directors that a further £200,000 was required, voted to wind up the company. "Contingent orders," Ward wrote, "were sent out by the August packet for winding up the concern, if on the arrival of their letters in Mexico, the mines should not be covering their own expenses."[11]

Ward's information may have been correct at the time he was writing, presumably in the summer of 1828, but in November of that year, a group of nine shareholders decided to resist the winding up of the company. By means of a formal requisition to the directors, they demanded a general meeting "to take into account recent favourable notices from the mines" and to reassess the closure resolutions taken at the previous meeting.[12] No account of the meeting has been found but it seems to have been held on 25 November. It is likely that the decision was to continue because three days later, the directors issued a further call for £10 a share and the shares were still being quoted early in December at £20.[13] Details of the Company's activities for the next few months are unknown but on 31 December 1830, a notice was inserted in *The London Gazette* summoning a general meeting of members for 1 February 1831 to arrange its dissolution. English stated in his *Quarterly Mining Review* (iv, 546) of December 1830 that the company's operations had stopped suddenly and improperly because of "mal-administration on the other side," presumably meaning in Mexico. £15 per share, he added, was returned to shareholders. Ward added a final comment. £320,000 had been spent by the company but in three years, no more than 33,501 dollars' worth of silver had been mined. "The establishment," he said, "is broken up; the gentlemen employed in it dispersed."[14]

The failure of the Catorce and Tlalpujahua companies was attributed at the time to a number of factors such as overextension of resources and personnel by acquiring too many mines; labor troubles in Mexico; technical problems with mining methods; shareholder impatience at the lack of prompt returns on their investments; and above all, financial difficulties as the whole mining experience in Mexico proved to be far more expensive than had been envisaged. The remaining five British companies—Anglo-Mexican, United Mexican, Real del Monte, Bolaños, and Mexican—survived for a further twenty years approximately but eventually with just one exception—United Mexican—they too succumbed to the same sort of problems. In addition, they had to contend with the effects of Mexico's more or less constant domestic political instability between 1830 and 1860; frequent changes in the laws affecting their activities in general and the mining industry in particular; and the effects of foreign intervention. In the end, by the late 1840s, the shareholders, some of whom had retained their shares from the beginning, had had enough and they refused to provide any further funds. The directors, reluctantly in some instances, were left with no choice but to close down all the operations, selling the mines to buyers where possible, and liquidating all the other assets in London and Mexico. With few exceptions, the first British experience of Mexico and its silver mining industry had been a very expensive failure. The reasons for the failure were, of course, many and varied but they may be conveniently summarized in the following areas: finance; labor; technical; environment.

FINANCE

Looking first at finance, all the companies were very quickly short of money. The millions of pounds raised in the market with the opening share issues were soon spent. By 1829, for example, the Anglo-Mexican Company had spent £1,032,663 or perhaps £73 million in modern terms. The United Mexican Company also raised and spent in excess of £1 million (£73 million). The other companies were equally free with their shareholders' money and with rumors repeated in the press of waste, inefficiency, general profligacy, and high management costs, complaints were soon heard at annual general meetings. *The Times* (30 August 1826) was especially critical of the Anglo-Mexican Company. Emphasizing that there was no question of fraud, the company, it alleged, was guilty of "extravagance, and an affectation of grandeur, as its besetting follies." Details published widely in the press seemed to confirm such accusations. *The Ipswich Journal* (17 February 1827), for example, included the following figures taken from a recent Company report:

Up to 31 August 1826, 3,300,000 pesos spent in Mexico
1,830 men employed in mines, even on the "reduced plan of operations."
146 women employed
40 English clerks and mechanics
1,949 horses in use
60 men employed to look after abandoned mines
301 men employed at reduction plants
1,184 mules in use

Others blamed the high wage rates for miners recruited in Cornwall and elsewhere and management salaries for the mine superintendents and commissioners were said to be far too generous. The costs of buying and transporting to Mexico hundred of tons of expensive equipment, especially the steam engines, had been underestimated. Even more costly had been the rush to contract literally hundreds of mines, some without even cursory inspection.[15] With unrealistic expectations of quick and high profits to be made, in many cases overgenerous terms had been given to the Mexican mine owners. The *alimentos*, or annual rental payable to the owners in return for the right to work the mines, were soon found to be excessive. The Bolaños Company, for example, contracted the Barranco mine at Bolaños for a twenty-five-year term at a fixed rent of 30,000 dollars, to be paid annually to the owner, José María Fagoaga who, incidentally, lived in France. Unable to meet such payments, many of the mines were abandoned or handed back to their Mexican owners as all the companies tried to cut back on their expenditure. By 1830, the United Mexican Company had returned thirty-nine mines and four refining haciendas on which almost £250,000 had been spent.

But heavy costs continued to be incurred or to escalate and the directors of all the companies had to appeal to their shareholders for new capital to keep the businesses going. Calls on shareholders for further funds became more and more frequent, but with prices of the mining stocks rapidly in decline once the market bubble had burst, those investors who had hoped to make a quick profit by buying and selling found themselves caught in the dilemma that always confronts speculators: sell at a loss or pay the calls. Many chose to forfeit their shares rather than pay the calls. For example, of the 10,000 shares issued and sold by the Mexican Company, 3,845 were forfeited by shareholders unwilling to pay the calls. The directors had to find alternative means of raising money. New share issues were regular events but again market hostility or disillusion with the mining industry usually meant that whole shares could not be sold. Instead, fractions of shares—half, quarter, eighth, sixteenth—were issued. Following its initial share offering in 1824, the Real del Monte Company, for example, tried to raise new capital no less than twelve times before the end of 1840. The Mexican Company was even more demanding, requiring payment by its shareholders of twenty-eight "installments" between 1825 and 1851. Some companies raised funds through interest-bearing loans repayable at short intervals. Diverse financial instruments including red and black debentures issued by the Real del Monte directors and miscellaneous securities and certificates by other companies were offered to shareholders, but all met with only limited success.

By the early 1830s, with no significant return on the huge investments that had been made, managers at some of the mines in Mexico were told that no further funds would be sent from London. Work at the mines had to be self-funding and if any were not making enough profit to cover expenses, they had to be abandoned or returned to their Mexican owners. Where promising new mines became available, and discoveries were constantly being made, they could only be acquired with profits made in Mexico. Regrettably from the British investors' point of view, several of the mines that were given up because they were unprofitable, promptly began to yield significant profits to their Mexican operators. "There seems to be a fatality," *The Morning Post*'s Mexican correspondent wrote in 1841, "attending all mines worked and then given up by English companies which begin to leave profits when in other hands."[16] His words turned out to be more prescient than he could have envisaged. According to Randall's calculations, the Real del Monte Company spent more than £3 million on its mines. Income was just over £2 million, leaving a loss of over £1 million. "Every penny," he concluded, "spent on a share in the firm's ownership was lost. Viewed through an accountant's eyes, the company's quest for riches among the silver-laden mountains of Mexico was a simple and total failure."[17] Ironically, however, within months of ending its operations and selling all its assets to Mexican investors in 1849, the company's mines at Pachuca enjoyed one of the most productive bonanzas of the nineteenth century. While some fortunate Britons who had shares in the new Mexican-owned company, for example, John Hitchcock Buchan, derived handsome dividends for the rest of their lives, the majority of the Real del Monte shareholders had missed what was almost literally, to coin a phrase, a golden opportunity.

There is no more graphic illustration of the financial scenario faced by the British companies than the record of their share prices (see table below).

Newspapers and contemporary writers often commented on this catastrophic collapse in share values, above all, in the Real del Monte Company. They showed, one observer wrote in 1843, "the madness of an English speculative spirit."[18] Investors had obviously made heavy losses and although in some of the companies, for example Real del Monte and Bolaños, many of the original shareholders remained remarkably loyal, retaining their belief that in the end persistence would pay off and fortunes would be made, there was a general reluctance to provide more capital. As a result, by the late 1830s, while the companies were still operating in several areas of Mexico, the scale of their work had been substantially reduced. In 1845, the Mexican

	1825	1831	1835	1840	1845	1849
Anglo-Mexican	135	29	6	1	2	1¼
Bolaños	545	170	130	55	6	3
Mexican	17	7	6	?	?	1
Real del Monte	1,300	45	20	2½	4	.63
United Mexican	180	10	5	2½	4	4

Company continued to work its mines in the state of Oaxaca where the profit for the year amounted to £2,296. A year later, a loss of £1,357 was reported.[19] Money on deposit in the company's bank account in London was £450. The Mexican Company had always been a much smaller and less ambitious enterprise than the other four British companies—it sent £361,789 to Mexico between 1825 and 1842—but they also had greatly reduced their activities. The United Mexican Company had raised £1,100,000 in capital and in 1848 it had £8,000 in the bank. One of the more successful companies, it had been obliged to give up its mines at Oaxaca, Zacatecas, and elsewhere, and by 1841 its activities were confined to a handful of mines in the Guanajuato district. It made profits, but only at best a few thousand pounds in the good years. In 1842, it made a loss of approximately £2,300. Its general manager in Mexico, son-in-law of William Bullock of the Egyptian Hall, James Nelson Shoolbred, on leave in London in 1845, was left in no doubt at a meeting of shareholders that he could expect no more money from Britain: "no engagement of any sort should be made upon the plan of requiring funds from this country, however tempting it might appear or promise to be."[20] Any plans for future developments, he was told, would have to be financed from funds raised within Mexico. Some shareholders thought that the company should be wound up, its assets liquidated and the proceeds distributed.

In fact, the financial picture of the British mining companies, gloomy as it was in retrospect to the thousands of investors who had risked their money in buying shares, was not all bad. Certainly, none of the companies had made any substantial profits or covered their huge investments, but some had managed to pay occasional dividends in the few good years that they enjoyed. The most successful in this respect was the Bolaños Company. In addition to the mines it acquired at Bolaños and elsewhere, it obtained the rights to 2,400 yards of the Vetagrande lode in Zacatecas. Its early experiences were not promising and having spent several hundred thousand pounds on the lode between 1826 and 1828 without any return, the directors decided to give up most of the concession. One small mine was retained. This was La Gallega and in January 1828, a bonanza was struck. Cross gives figures to illustrate the impact of the discovery.[21] In the first quarter of 1828, 194 tons of ore were mined. By the fourth quarter, the figure rose to 3,680 tons. Profits from the mine enabled the company to invest heavily in other mines and between 1829 and 1836 silver to the value of 2,200,000 pesos per year was produced, giving in 1830 alone almost 1,000,000 pesos profit. By 1838, the bonanza was over but the company's shareholders had received annual dividends for six years. These were not notably large, ranging from 7.3 percent of the full value of a share in 1830 to 2.7 percent in 1833, with payments in between these figures in the other years. 1836 was the last time that Bolaños shareholders received anything. Thereafter, the company declined and being unable to raise further capital, it disposed of its mines and began to wind down its operations.

The Bolaños shareholders were not the only ones to receive some return on their investments. The United Mexican Company also paid dividends, usually at the rate of little more than a few shillings per share, although in 1841 holders of some of the company's new shares were paid £1 per share. Likewise the Mexican Company was able to pay the occasional dividend—7s. 6d., for example, in 1845. The Real del Monte investors, many of whom, as we will see in another chapter, remained loyal to the company throughout its existence, were less fortunate. Frequent additional demands on their pockets were made by the directors but no dividends were declared and those who had bought their shares in the early years when prices were caught up in the stock market bubble suffered particularly heavy losses. The Anglo-Mexican Company, with the largest capitalization of all the British companies at well over a million pounds, was just as bad. Not a single dividend was declared in its twenty-five-year life. Its long-suffering shareholders had to wait until the company was dissolved and its assets sold before they received any return on their investments.

LABOR RELATIONS

Lack of capital was by no means the only difficulty confronting the British mining companies. In the field of labor relations they experienced many problems arising from a combination of ignorance of local custom and practice; arrogance and overconfidence in the superiority of British mining knowledge and technology; and complications arising from the occasionally unruly character of Cornishmen and other Englishmen abroad.

When the companies arrived in Mexico in the mid-1820s, they brought with them a managerial class that included some individuals with a sound knowledge of mining but also retired army officers and very young men who were new to the industry. Few, it appears, had any knowledge or understanding of Mexican mining practices and they were not impressed with what they found on their arrival in the country. These words by Justus Von Uslar, chief director of mines in Oaxaca for the Mexican Company, offer a good summary of the patronizing attitudes that he and other Europeans adopted:

> As Miners, they can be rendered very serviceable, so long as they are kept to the use of Hammer and Pick: as Blasters, and still more as Timber, or Wood-workmen, they are below mediocrity, and altogether destitute of mining knowledge. Notwithstanding, they already begin to look on, and understand our better methods; and, in time, I am not without hope that they will make considerable progress in improvement.[22]

Uslar, and many others like him, firmly believed that however Mexicans and Spaniards before them had managed to extract precious metals from the earth for the previous three centuries, European techniques and technology were superior. Above all, it was believed that the major problem of flooding in the mines could easily be resolved with the use of powerful Cornish beam engines.

The introduction of new technology and new practices quickly caused problems with the Mexican workforce that was employed in very large numbers—several thousand men—and conflicts in the form of strikes and occasional incidents of serious violence soon ensued. The principal cause of conflict involved the *partido*, or Mexican practice whereby the miners or *barreteros* were allowed to keep a share of the ore that they had mined. They sold their share of the ore back to the company or to independent contractors and the money they earned was a very important addition to their regular wages. It was also a significant cost to the companies and most, if not all, the British managers attempted either to abolish *partidos*, to reduce them, or to restrict them in one way or another. Several attempts were made to replace the *partido* with Cornish type tut-work contracts whereby groups of miners were contracted to work on an incentive or payment by results system at previously agreed rates. All such innovations were fiercely resisted by the Mexican miners, sometimes with violence against the British managers or against Mexican army units brought in to keep the peace in the always volatile mining camps. Both Todd and Randall give many examples of incidents at Pachuca and Real del Monte. Starting in 1827 with a strike by the *barreteros* over proposed changes to the *partido*, the Real del Monte Company faced regular disruptions—summarized by Randall as "strikes, uprising, skirmish and dispute"—with its workforce over the next twenty years.[23]

Militancy in defense of traditional working practices was not confined to Real del Monte. Similar conflicts over the *partido* and other changes that the British managers tried to introduce took place in Zacatecas, Guanajuato, and elsewhere. A serious situation occurred at Zacatecas in April 1828 when the miners' demands for an increase in their *partido* were refused. Riots followed and the British managers were forced to take refuge in a building "where they were assaulted by a large mob." Negotiations with miners' representatives calmed the situation for a time but when no agreement was reached, the mob, estimated at 1,000 to 1,500 people, returned to attack the British house with stones and other missiles. The occupants, no doubt fearful for their lives, opened fire but we have no details of any casualties. The crowd dispersed but then set fire to and destroyed company-owned buildings at the mines, causing damage estimated at 85,000 pesos.[24] Lucas Alamán reported to the directors of the United Mexican Company in 1830 that the workmen at the San Aacasio mine in Zacatecas were refusing to work "without being allowed an increased *partido*."[25] Disturbances were reported at

Tlalpujahua in 1829. An alarming state of affairs was said to prevail with "riotous conduct of the miners who are attempting by violence and intimidation to dictate to the Company the terms by which they are to be remunerated for their labor.[26]

Disputes over working conditions with the Mexican miners were not the only problem. All the companies in every mining area of Mexico experienced labor shortages. There was a variety of causes. Mexico's chronic political instability was often accompanied by military rebellion and forced conscription from local populations was common, either by incumbent authorities seeking to defend their position, or in areas where rebel army officers gained control, albeit temporarily. When news of an approaching rebel force was known, men deserted the mines in large numbers, fleeing to outlying districts where they could disappear until the military intruders had departed. The Anglo-Mexican Company's manager in Guanajuato reported that "In consequence of the unsettled state of affairs, nearly all the young and able workmen had absconded, to avoid being pressed for the army." On another occasion, the manager wrote that Guanajuato was threatened by "the approach of hostile detachments (the partisans of General Guerrero). The impressments of men for the military service, which was in consequence resorted to, kept many away from their work."[27] Fleeing the press gangs in times of rebellion was common but in times of national emergency, for example, the war against the United States from 1846 to 1848, young men from the mining communities volunteered to join the army.

Another problem, unanticipated and rarely fully appreciated by the British, related to the attitude of at least some Mexicans to work. Here there was a paradoxical situation for the British. All the companies required very large numbers of animals—mules and horses for the most part—to operate the grinding and pumping machinery used in those mines which did not have beam engines or where the more modern technology could not be used. The bill for maize used as fodder for the animals was one of the major items of expenditure and the mine managers were always alert to the state of the market, buying in bulk when prices were low and keeping large stores in reserve. Market prices were determined by supply and demand and when harvests were poor, prices rose and when they were good they fell. The Mexican workman and his family had maize as their staple diet and they were even more directly affected by prices than their British managers. The British found that in times of shortages, the Mexican miners were more anxious to work to earn extra money to compensate for higher prices of their basic food. In contrast, when prices were low following good harvests, the workers were less inclined to work. Referring to the "deficient supply of labour," one manager explained:

> The circumstance of the article of maize (the chief food of the working classes in Guanajuato), having, in consequence of two abundant harvests in succession, become extremely cheap, has had a most disadvantageous effect on the industry of the working population, leading them frequently to work only two days in the week.[28]

It was also the case, however, as was noted in Zacatecas in 1831, that:

> The low price of provisions, the season having been particularly favourable, was the principal cause of the deficiency of hands, an inconvenience in some measure, counterbalanced by the cheapness of agricultural supplies to the mines and haciendas.[29]

The negative impact on productivity arising from the Mexicans' predilection not to go to work if there was food already on the table, was compounded by the national calendar of secular and religious holidays when work was either forbidden or not required. The religious calendar attracted much hostile comment from the British managers because there were so many "holy days" when the Church authorities expected their congregations not to work. According to Henry Ward, "Feast and fast days reduced the number of *días útiles* (on which labour is permitted) to about one hundred and seventy-five."[30] Whether Ward's figure is accurate is not known but certainly even the clergy recognized that the number of religious festivals was excessive. Some changes were made in later years but the religious calendar remained a problem for the mining companies. The secular calendar was also increasingly crowded as the republican governments sought to create the nation's own pantheon of heroes. Independence day celebrations on 16 and 17 September and sometimes on the 27th and 28th of that month were the main public holidays but others were added from time to time. The British companies tried to get around the problem of public holidays. They included in their employment contracts a clause that their workers would only be excused work on those days in the calendar that were public holidays in Britain. Whether the workforce took any notice of this instruction is not known.

Another difficulty encountered by the British was the Mexican worker's custom or habit of helping himself to a little extra reward for his labors. In a report in 1826 to the Anglo-Mexican Company, the managers at Guanajuato informed their directors in London that, "the *Buscones* continue refractory and did not descend last week, owing to our having insisted on selling the ores by public auction. The former system of dividing the ores themselves we found gave rise to much pilfering."[31] Theft of ore or silver was always a problem in every area and Todd provides details of some of the more inventive methods used by the Mexican workers at Real del Monte to escape the attention of the guards, supervisors, and others who watched, at least in theory, their every move:

> They crushed the silver ore into powder, mixed it with tallow and rubbed it as a pomatum into their bushy hair. They hollowed out the handles of their hammers and filled them with pieces of pure silver. They concealed the grains of silver and gold in their mouths, teeth, ears or any other kinds of orifice they chose.[32]

One especially gruesome incident related by Todd involved the corpse of an overseer killed in a mining accident. The miners disemboweled the body and filled the empty stomach with ore.

None of the problems so far outlined compared in significance to the scarcity of labor caused by disease. Mexico endured regular epidemics of smallpox, typhoid, cholera, dysentery, and other illnesses that decimated the population in some areas and hit the mining communities very hard. Smallpox, for example, struck Guanajuato in 1830 and at the Valencia mine worked by the Anglo-Mexican Company, the manager reported that during the months of June and July "the number of sick at the Valencia mine alone averaged from 100-150 per diem, and the deaths from 25 to 35 per diem."[33] "Sickness increased, chiefly smallpox," he noted, with "150 on the doctor's list, many deaths, chiefly children and young people." Fortunately, he added, none of the English residents had been affected "but they had been much visited by dysentery."[34]

The apparent immunity of the English and other Europeans to some diseases was attributed by one English manager to a good bottle of wine. Two of his German employees were sick and confined to bed with a "complaint" or "species of fever which can only be cured by wine." He continued:

> Myself and suite enjoy an uninterrupted state of good health, and it is my opinion, and that also of the medical men who attend us, that our using wine at our meals is the reason why our health is so good as it is. I beg leave, therefore, to suggest that the Board would authorize me to supply to the six men wine at the rate of half a bottle per day, which costs us about a dollar a bottle. This would be an economical arrangement, for independent of medical charges, the men's time would be saved, and the works would not suffer those hindrances, the constant ill health of the men subject them to.[35]

We do not know if the Board of Directors approved this request but wine was certainly no defense against cholera. In 1833, a major epidemic ravaged several areas of the country, killing many Mexicans and also an unknown number of Europeans. At Bolaños, for example, it was reported that over a third, or nine of the twenty-five English working there had died.[36] Oaxaca, Mexico City, and elsewhere were struck in later years and Britons, including William Bullock's daughter Sybylla and son-in-law, James Nelson Shoolbred, were among the many victims.

The use or abuse of alcohol was an area of much concern to the directors of all the companies. When the various companies were established in the 1820s, it was assumed that the bulk of the labor force would be drawn from the local population, especially in those regions with a long tradition of mining like Guanajuato and Zacatecas. It was also recognized that there was a need for skilled miners familiar with the industry and as we have noted earlier, many of these were drawn from Cornish mining communities. There are no comprehensive statistics of the number of men who left their Cornish villages and headed to Mexico, attracted by the generous wages on offer, but there were certainly several hundred. Many found it difficult to settle and turned to alcohol for relief. Drunkenness became common in all the mining areas and there was much hostile and critical comment on the behavior of the Cornish miners. Captain James Vetch, the Real del Monte commissioner, was one of the first to complain:

> I must here express the disappointment with the miners sent me from Cornwall . . . Many of these men disgrace the streets with their drunken brawls at night and will not work unless goaded to it and seem perfectly indifferent to the needs of the Company.[37]

A decade later, excessive drinking was still a problem. Todd cites the case of Titus Geach, unceremoniously sent home on 12 June 1835 for what his employer described as "your habitual drunkenness, the bad example you show, and the disgrace you bring upon the English character."[38]

Of course, recruits from Britain and elsewhere in Europe, most notably Germany, were not confined to underground workers and experienced miners. A whole range of other skills was required. There were chemists, doctors, carpenters, masons, blacksmiths, clerks, accountants, and all manner of other trades. All the London-based managers who carried out the recruitment in Britain went to considerable lengths to ensure that the personnel they hired were suited to the demands of working abroad. Long and very detailed written contracts were drawn up, even for the most humble of positions. A good example is the case of Joseph Owen. He was recruited in London in 1834 by John Taylor, the British-based manager of the Real del Monte Company. Very unusually, his contract of employment has survived.[39] It consists of three large and very closely printed pages of text in which the terms of his employment are spelled out in detail. He was required to obey all orders given to him in Mexico; he must work full-time for the company and not accept any other employment; he must work, if required, on days that were public holidays in Mexico but not in Britain; he must give three months' notice if he wished to terminate his employment. His salary was 480 dollars (almost £100) per annum of which the Company was to advance him £30 for outfit and other expenses and the Company was to pay his travel expenses.

At the end of the contract, there is recorded an additional request that Owen made, and to which the Company agreed, that £2 10s. a month should be paid from his salary to Mary Owen, of Woolwich (presumably his wife or mother but it is not stated). Finally, there was the following clause:

> The said Joseph Owen shall pay due and becoming respect to the Civil and Religious Constitutions and Customs of Mexico, and shall carefully abstain from giving cause of offence in relation thereto, and from all interference whatever therewith, or with the Politics, or Government of the Country.

We know from the Mexican passport records that Owen arrived at Veracruz in January 1835. His contract was for three years' work, and at the end of that time, he resigned. His movements thereafter are not known but by the summer of 1838, he was in Mexico City and he was destitute. He applied to the British consulate for relief and for help in paying his passage to Veracruz and to Britain. As far as we know, he was given assistance and returned home.

Owen's case is a useful and in many respects a typical example of the British workers. Many of the Cornish and other Britons who traveled to Mexico to work for the mining companies did not stay in one place for very long. As Todd put it, "the miner and his family moved around as the whim, the ore and the pay affected him."[40] Once their initial three-year contract was completed, they either returned home, or went to work for companies in other parts of Mexico, or left the mining industry altogether to seek other ways of making a living. Some disappeared from view altogether and the British diplomatic records have many cases of mothers or wives asking for help in locating their missing sons or spouses. Experienced and skilled miners were always in short supply and the British companies were obliged to send new recruits from Britain every few years, paying high wages and other benefits to attract them. Wage costs spiraled and as expenditure in the 1830s onward began regularly to exceed income, the managers attempted to reduce wages, generating even more industrial strife.

Finally, in a few places, labor shortages were ameliorated by the use of convict labor. In Zacatecas, for example, the mines at Fresnillo had the use of about 100 convicts a day for which the companies paid the jail the going rate for laborers. The men were escorted daily to the mines by a military unit and once there they were "available for employment in the severest labours."[41]

TECHNICAL PROBLEMS

The British mining engineers did not travel to Mexico in the 1820s entirely ignorant of the practical problems they were going to face. They were well aware from their familiarity with Humboldt's works or with many other sources that their biggest challenge was going to be water. Most of Mexico's mines had been abandoned years before and as a result they were flooded. Underground galleries, shafts, and drainage adits, some hundreds or even thousands of feet in depth or length, were filled with water. An indication of the scale of the work needed at the Real del Monte mines, for example, was noted by one of the young British managers, John Hitchcock Buchan. Before the miners could start work, he wrote, they had the task of:

> clearing out an adit upwards of a mile long, and restoring numerous shafts (generally found filled up, or in a most ruinous state) to a depth of 110 fathoms, before they could even commence the drainage of the lower part of the mine by their steam engines.[42]

Removing the water was clearly an essential prerequisite before any mining could begin. There were two options. The first was to use the well-tried Mexican method, developed over many years, of using what were called *malacates*. These were winding machines moved by horse power and not unlike the whims once used in British mines. Four horses were used to work each *malacate*. They raised to the surface water collected in bags made of raw ox hide known as *botas*, each one when full containing approximately 750 pounds of water. The horses were only worked for two hours in each twenty-four which meant that forty-eight horses were needed daily for each machine plus two that were kept in reserve. Ten men were also needed to operate each *malacate*. The cost was obviously considerable, not just the men's wages, but even more significant the cost of the horses and forage for them. One estimate put the weekly cost at 165 pesos "when forage is at a moderate price," but in years of poor harvests the costs rose. Each *malacate* was operated on a twenty-four-hour basis, but the number needed in a mine depended on the season, more being required in the rainy season and less in the dry. Even so, as many as thirty-nine were being used at the Fresnillo mines in Zacatecas in 1833, using 1,950 horses and 390 men.

The *malacates* were generally recognized by the British as reasonably efficient, managing to raise steadily vast quantities of water, but they were very costly. They did not compare, or so it was believed in the early years, with the power and efficiency of pumps driven by Cornish beam engines. Hence, all the British companies invested very heavily, spending many thousands of pounds on large and expensive engines. Heroic feats were performed by engineers and their assistants in transporting the machinery in parts across the Atlantic, followed by very hazardous and extremely difficult conditions on the dirt roads of Mexico, at times across high mountain ranges and inhospitable climates, and then after months of exhausting travel, successfully reassembling the machines on site. The following example from the

Bolaños Company records offers a good illustration of the various problems and costs involved:

> The last report noticed the progress of the transport of the steam-engine intended for the drainage of the Barranco mine, and it was stated that it would probably reach its destination by the middle of March, 1830. The engine appears in fact to have arrived at Bolaños on the 18th March, the time consumed having been 108 days, viz. 77 days of actual travelling, 10 days spent at Perote altering the harness, repairing wheels etc. and 21 days' delay at Appan, while awaiting the result of the negotiations as to the Barranco contract. The distance is 286 leagues, the weight conveyed was 39 tons, and the expense, including wear and tear of wagons, harness and cattle, but exclusive of the delay at Appan above mentioned, 16,800 dollars, being 500 dollars per ton, or rather less than two dollars per ton per league.[43]

Unfortunately, from the British investors' viewpoint, successfully financing, transporting, and installing engines at the mines was not all that was required. Several problems ensued. The first of these was a source of fuel to drive the steam-powered engines. Most of the companies, at least in the early years, hoped to be able to locate and develop coal mines in Mexico, but despite much exploration and raised hopes on many occasions, no significant or usable coal deposits were located. The alternative was to use wood, and vast areas of forestry were acquired from Mexican owners. The Bolaños Company, for example, secured a twenty-one-year lease at 1,500 pesos a year on the so-called "Forest of Camotlan." The Company's surveyors reported that the estate:

> was found to run in a northerly direction for about thirty miles, and the same distance east and west, embracing, therefore, a surface of about 900 square miles. Of these there are about 14 square miles of good arable land in patches of from 10 to 80 acres each, and 200 acres in small patches of from 2 to 6 acres each, that may be irrigated and adapted to the cultivation of maize, wheat, all the artificial grasses, and most vegetables; the remainder of the estate consists of beautiful pine, oak, and a great variety of other useful woods.[44]

In some instances, while the mineral deposits showed promise, the absence of readily available timber supplies meant that steam power could not be used. In those cases, the only alternative was usually to return to the *malacates* with their own particular demands for men and horses. In other areas where rivers were near the mines, it was sometimes thought possible to harness the water supply as an alternative to steam power and there were impressive feats of engineering attempted. At Bolaños there was "a copious stream of water, capable of being applied to very powerful water wheels for draining." The British engineers, therefore, decided to make use of the river:

> It was proposed to put up an over-shot water wheel, of seven feet in breast, with an effective fall of forty-two feet, the river being capable of affording a supply in the driest season of thirty cubical feet per second, equal to a quantity weighing 112,000 pounds in each minute; to obtain, however, this water-power, it was necessary to conduct the stream by a long and expensive leat or water-course to the place fixed upon for the erection of the wheel.[45]

Another example was in Oaxaca where the United Mexican Company worked the Natividad mine. The reduction hacienda used for the ore taken from the mine had its machinery consisting of two stamping mills of twelve stamp heads each "moved entirely by water.[46]

The use of what was then a relatively new technology in the form of steam power also required many skilled men, familiar not only with the operation of the engines but also how to repair them in the event of break-down. Blacksmiths were always in high demand but when engine parts failed, it was often found that there was no iron available to either repair or replace them. Anticipating the problem, several companies acquired known iron deposits and invested heavily in developing mines in those areas. The United Mexican Company, for example, opened iron works at Durango and put a German engineer, Mr. Seyffert, in charge. He built blast furnaces, dammed a river to create a water mill to operate them, acquired some charcoal and iron ore, and was able to produce a small amount of rough iron. He hoped, he told his directors, "that many articles of use at the mines, might be cast." His hopes did not come to fruition and the Durango ironworks were closed shortly afterward.[47] None of the other attempts to produce iron had any quick success and repairs had to rely on the ingenuity of the blacksmiths or on the long delays in obtaining spare parts from Britain.

Draining and clearing the mines to gain access to the ore was, of course, only the first stage. Once the ore had been raised to the surface, the silver presumed to be within it had to be extracted, and it was in the extraction processes that the British miners encountered major problems. Once again, at first, they had little respect for the traditional and well-tried Mexican method. This was the patio or amalgamation process. Much has been written about this elsewhere and only a brief description of what was a very complex process is given here. The ore was first broken into small pieces with hammers and then taken to stamping mills where it was ground to a fine powder. This was then heaped onto a paved court or patio at the reduction hacienda. Water and various chemicals were then added to the mix until the *torta* as it was known, or heap of ore, began to heat up and chemical reactions started. Lime and quicksilver or mercury were sprinkled onto it and an amalgam was formed. Eventually, after numerous washing and mixing processes, the mercury was volatized by heat and some of it collected when it condensed in specially prepared pipes and cisterns. Finally, by a process of washing and applied heat, the silver content of the *torta* was recovered. It was poured into a mold and cast into bars that were normally taken to a mint for coinage.

The whole process took several weeks and required the alert and experienced eye of the *azoguero* or amalgamation expert who was able to judge when to apply the chemicals, and especially, the mercury. It was not a particularly effective way of extracting silver, and sometimes gold, from the ore. A high percentage of the precious metals was lost and the whole process was dependent on an assured and affordable supply of mercury. Ensuring the supply of that essential ingredient was always a problem for the British companies. For hundreds of years, the world's supply of mercury had largely come from Almaden in southern Spain. Then, in 1831, the Rothschilds gained control of Almaden and several other minor sources around the world, thus creating a monopoly. Prices rose accordingly, from 90 pesos a *quintal* (100 lbs English) in 1833, to 135 pesos in 1837. The mining companies had to seek efficiencies or reduced operating costs to pay the rising prices but they also went to considerable lengths and expense in a search for local or Mexican supplies. Several promising mercury mines were discovered in Chihuahua and elsewhere. Work was carried out in them but it was not until the late 1840s when vast deposits were found in California at New Almaden that the Rothschild monopoly was broken and prices began to fall.

Smelting was the preferred Cornish or British method of treating the ore and it was tried in various ways by several of the companies. None achieved any great success and it was usually found that the acquisition and price of fuels like timber and charcoal were not cost effective when compared to the patio process. All manner of other new methods were tried, such as the Freiburg or barrel amalgamation process; the flotation method; and the Spangenberg system.[48] Many experiments were also conducted in the search for an improved refining process and various applications were made to the Mexican government for patent protection of what the applicants believed to be a new and efficient process. To give just one example, in February 1839, Charles Penny, an Englishman, described as a miner in the state of Sinaloa, petitioned the government for a "life-time privilege" for his new way of refining metals.[49] There were many similar petitions from British and other nationalities among the mining community but none of the new methods seems to have improved on the patio process.

ENVIRONMENT

Any cursory survey of the literature generated by the British mining companies—directors' reports to shareholders; managers' reports from Mexico; private correspondence; diplomatic summaries; letters to the press—reveals one very prominent feature. The first generation of British citizens to live in independent Mexico complained, not just about the food and climate, or customs and culture, but about almost every aspect of Mexican life. Miners and merchants complained constantly about government legislation that affected their interests. They objected to changes in the customs tariff, to laws on taxation, to decisions on imports and exports, some favoring free trade and others protectionism. They appealed with regularity to British diplomats for help in avoiding or evading the effects of legislation or for perceived injustice on the part of the Mexican authorities. The record, perhaps, for persistence and number of written arguments in pursuit of a personal claim against Mexico was held by Robert Kinder. Closely concerned as we have seen in the foundation of the Real del Monte and Bolaños Companies, Kinder was also involved in trading and financial investments. Without going into the details, it was in connection with these that he pursued his case through the courts in Britain and in Mexico. He also badgered the British Foreign Office with often very long expositions, setting out his claim. In May 1837, to demonstrate how much time and effort had been expended on the matter, an official compiled a list of seventy-four documents that had been accumulated in the Kinder file.[50] Although he was promised and probably given informal support by British diplomats in Mexico City, he was not given full diplomatic backing for his grievance against the Mexican government. The reason given to him was the same as that given to many other British complainants. Unless the complaint arose from an area covered by convention or treaty between the two countries, the British government would not provide support. From its viewpoint, the complaint or case was a private one and, therefore, not a matter for Her Majesty's Government.

Some of the British demands—complaints is too mild a word for many of the cases—dragged on for decades, being pursued through the courts and corridors of power on both sides of the Atlantic. Kinder's case was strictly individual but there were numerous other examples of demands/complaints/protests arising from the activities of the British mining companies. Two examples will suffice. The first concerned the United Mexican Company. Mexico's chronic domestic political instability with frequent military coups and violence was at least in part the result of the weakness of federal/central and state authorities and their inability to control rebel army officers and dissident military units. *Pronunciamientos*, or rebellions, occurred with sometimes bewildering regularity and it became a commonplace of the time that they were an easy route for army officers to gain promotion or for seriously underpaid and ill-equipped men in the ranks to make some money to boost their often non-paid rations. One such rebellion took place in 1833 as part of what became a series of similar events during a particularly unstable period. I have discussed it elsewhere and here we need only note a few details.[51] The rebel leader was General Mariano Arista. He declared himself and his men in revolt in June 1833 and after spending some weeks in the south of the country, he was forced to change tack and head north. At the head of a bedraggled force of men and some women, he passed through the mining areas of Pachuca and Real del Monte, followed by his hometown of San Luis Potosí. Then he led his followers north to Guanajuato. In political or military terms, there was no logic to this move and the assumption at the time was that he went to the famous silver mining city because he, and his men, knew that there would be booty to be had from the silver mines.

There were, of course, many mines and companies in Guanajuato but among the most prominent, and apparently best funded in 1833, was the United Mexican Company. Arista lost no time in confiscating the company's cash and other assets. His men helped themselves to the stores of fodder accumulated for the company's livestock and piles of ore found at the mines were sold in public auction to local refiners. The homes of the resident British managers were occupied by Arista's officers and the company's offices were looted. Loud and angry appeals for help were made in vain to the local judiciary and mining authorities and even President Santa Anna refused to listen, arguing that neither he nor his government could be held responsible for the actions of rebels.

The Arista rebellion ended in October 1833 when Santa Anna and his army occupied Guanajuato. Arista was sent into exile in the United States, allegedly taking with him the cash he had stolen from the British. The value of the money and goods taken by him and his men was around 85,000 pesos and the United Mexican Company began what was to be a long and expensive legal battle for compensation. Court cases were initiated, reports were drafted, appeals to national governments were made, directors in London went to the Foreign Office to put their case for help, and diplomatic assistance was provided by Richard Pakenham, Britain's senior diplomat in Mexico City. For years the pressure brought no success and although eventually some compensation payments were made, the Company failed to get what it considered proper and full payment of its claim.

The United Mexican Company was also involved in another protracted case that involved repeated, but unsuccessful judicial and diplomatic intervention. Contractual disputes with the owners of the mines that had been leased by the British companies were common and much to the anger of the British managers who tried to resolve them in the courts, they almost invariably found that Mexico's judicial system relied more on bribery and corruption than honest appraisal of a dispute. What became known as the San Acasio claim concerned mines that the United Mexican Company was working at Zacatecas. In March 1833, the Mexican owners alleged that the British miners were destroying the mines through neglect and incompetence. They persuaded the local magistrate or *alcalde* to issue a sequestration order on all the United Mexican Company's property. Horses, tools, gunpowder, straw, corn, and substantial amounts of cash were seized and handed over to the Mexican owners. As they had done in Guanajuato when Arista had seized their property, the British managers protested as best they could. They attempted to use the judicial process at local, state, and federal levels, but all to no avail. One local lawyer employed by the company to make their case concluded that "the whole business has been a conspiracy, authorized and supported by judicial formulas, one more illegal than the other, against the company which worked the mine."[52] Pressure by the directors in London on the Foreign Office again persuaded the British government to instruct Richard Pakenham to take up their case. He pressed and pressed for years but as in the Arista case, with only limited success.

The Arista and San Acasio claims were supported, albeit without much effect, by the British Foreign Office but there were many other cases in which diplomatic help was refused. During Mexico's numerous rebellions and other occasions of widespread civil disorder the British mining companies suffered losses to both rebel and government forces as law and order broke down. Forced loans were common, theft of property frequent, and damage to mines and installations almost a regular event. In 1829, for example, a gang of thieves plundered the United Mexican Company's property at Sombrerete in Zacatecas. Details of the incident and demands from the company's directors for help in seeking compensation from the Mexican authorities soon reached London. The Foreign Office refused to help on the grounds that "the loss appears to have been occasioned by a violent outrage committed by a band of robbers who do not seem to have possessed any military character to give the Government an implied control over their proceedings."[53] Protection was also often sought from the Mexican authorities. The silver convoys with their very valuable cargoes were always at risk on their journeys from the mines to the ports at Tampico, Veracruz, or Mazatlán. Military escorts were essential and were provided by the Mexican army. Sometimes, military support was requested to keep the peace and to protect the British workers at the mines.

Another more unpleasant and unexpected feature of the environment in which the Cornish and other miners had to work, at least in the early years, was the attitude of the Mexican people toward the British companies and their staff. When the companies were first formed in London and their personnel recruited, assurances had been given by Mexican representatives such as Lucas Alamán that the British would be welcomed in all mining regions of the country. In the very early weeks and months after their arrival, that was probably true, but it is also the case that within a short time there were increasing signs of popular hostility toward them. We have already mentioned the labor disputes that took place as the British tried to impose change on local working practices. Other difficulties were soon apparent. In the summer of 1825, a pamphlet appeared on the streets of Mexico City with the title "O matan a los ingleses o el Gobierno los proteje (Either kill the English or let the Government protect them.)" The anonymous writer repeated stories such as the murder of an Englishman who had refused to kneel before a religious procession. A visitor recorded that "in the eyes of a Mexican, every foreigner is English, and every Englishman is a heretic."[54] Some of the British newcomers experienced difficulties because of their religion. For example, William Bullock's daughter Sybilla and her fiancé, James Shoolbred, could not find a priest in Mexico City who was willing to marry them because Shoolbred was not a Roman Catholic. The couple eventually found a French priest willing to conduct the service. There were problems over burial grounds with British Protestants being denied interment in Roman Catholic consecrated ground. After much delay and debate, compromise was usually found with Protestant or non–Roman Catholic cemeteries being allowed outside the towns as, for example, the English cemetery at Real del Monte. Personal safety was always a worry. Francis Baring told his father that, "We travel here armed up to the teeth and with reason for all parts of the country are infested."[55] Even Richard Pakenham found that his diplomatic status offered no protection. Soon after arriving in Mexico in April 1827, he wrote in a private letter to a friend that, "I have already been once robbed and narrowly escaped being murdered. It bores me exceedingly to be obliged to go about armed always as if I was going on a crusade to the Holy Land."[56]

Pro-Spanish conspiracies were discovered in 1827 and political parties began to campaign for the expulsion of Spaniards from the republic. The anti-Spanish campaign was easily extended to all foreigners and random attacks on them increased. News of the situation in the country was repeated in the British press and the formerly positive image was changed. A good example are the words of George Borrow, the noted traveler and evangelist. His brother, John, was in Guanajuato in 1833 working for a British company when Arista ransacked the city.[57] The events were reported in the British newspapers and George was concerned for his brother's safety. Writing from Russia to a friend in London, he wrote of "the late dreadful commotions in Mexico, that most benighted and miserable region." He had read in the papers of how the town of Guanajuato had been attacked by "murderous bands of insurgents." They had been "incited by their demoniacal priests, in comparison with whom the Shamans of Manjuria and the Lamas of Mongolia and China are innocent and holy, to lay hold of every opportunity of shedding the blood of Protestants and foreigners."[58]

Borrow's fears were exaggerated but not long after cries of "mueran los ingleses" ("death to the English") were heard again in the streets of the capital. Even in congress what was said to be harsh and intemperate language was used against the British. There were other incidents of anti-British feeling in Mexico City and in some of the provincial towns but in general, after the mid-1830s, the local population in the mining regions accepted their visitors with less hostility. Mutual tolerance prevailed.

END OF THE COMPANIES

It was almost certainly no coincidence that of the five surviving British mining companies established in the 1820s with such fervent hopes, four were wound up within the space of three years from 1849 to 1852. Their shareholders had endured too many disappointments and were unwilling to provide more capital. There were sound reasons for their disillusion. The stock market panic of 1847, with the bursting of the railway shares bubble, had made investors more cautious. The next year, 1848, the year of revolutions across Western Europe, had been no more encouraging. Across the Atlantic, war against the United States from 1846 to 1848 had cost Mexico almost half of its national territory and while the military events do not appear to have had much direct effect on the mining industry, the devastating blow to Mexican prestige and economic potential was not lost on British investors. Furthermore, other avenues for investment were opening up to those interested in mining. In California, the discovery of gold in 1848 prompted a huge influx of men and money, not unlike what had occurred in the 1820s for Mexico. Joint stock companies were quickly formed in London to exploit the Californian bonanza and thousands of new shares were offered for sale to the British public. Australia also offered rapidly expanding opportunities in the mining and agricultural sectors and again, shares were on sale from several new companies. Mindful of what had happened in 1825, the public was strongly advised by the press to be wary of the promises made in the new company prospectuses:

"It is to be hoped that Englishmen, warned by the lessons of a bitter experience, will not, on any large scale, repeat the madness of an epoch so disgraceful in our commercial history."[59]

Faced with such reminders of the past and what had happened in the 1820s, investors were in no mood to risk more of their capital. Four of the British companies were wound up and all their few remaining assets liquidated. The first was the Real del Monte Company. The most famous of all the British mining companies because of the extraordinary rise in its share price to approximately £1,300 for a £400 paid-up share, it had been struggling for a number of years. At a shareholders' meeting on 24 June 1848, the chairman, Sir Robert Price, painted a stark and gloomy picture of the company's prospects.[60] Recent heavy investments in new machinery and new methods of refining poor grade ore had failed to yield the hoped-for returns. Losses were being incurred on the mining operations and more capital from Britain was needed if the company was to survive, but as it was recognized that there was little prospect of raising additional funds, work on most mines had been suspended. During the next few weeks, more reports arrived from the general manager at Real del Monte. This was John Hitchcock Buchan. We have already mentioned him and many more details of his career will be given in other parts of this work. All we need to note now is that he had been one of the company's first employees, arriving at Real del Monte in 1825. He worked there and elsewhere until 1836, when he left to investigate the discovery of new silver deposits at Guadalupe y Calvo in the mountains of Chihuahua. He returned to the Real del Monte Company in the summer of 1848 to find that there were major problems. He immediately introduced economies "in every branch of the establishment," but losses continued at the rate of 2,000 dollars a week. Additional capital, he told the directors, was vital if the enterprise was to survive.

Sir Robert Price relayed Buchan's opinion to the assembled shareholders at another meeting on 28 August 1848.[61] The Board of Directors had concluded, he said, that it was not possible to raise any more money. Their opinion was that it was time to close the company, sell off its assets, both in Britain and Mexico, and to pay its debts to loan-note holders and others. In reply to a question from the floor, he stated that the value of the works, exclusive of the engines, was £80,000 but that there was an outstanding debt in Mexico of £15,000. One shareholder, Mr. Heath, objected to the proposed sale, arguing that extra funds could be raised and that "the mines would pay ultimately, especially as hopes were held out of discovering a rich course of ore." A director, Horace Twiss, disagreed, insisting that no new funds were available. After another two hours discussion, the meeting was adjourned "to allow the proprietors time to consider the subject."

A month later, on 25 September 1848, at a specially convened meeting, Sir Robert Price reported that "the directors had come to the conclusion that it was imperatively necessary for the safety of all concerned in the undertaking that the company should be dissolved immediately."[62] After a long discussion, the resolution dissolving the company was carried. The reason for the urgency was subsequently revealed to have been another message from Buchan in which he warned that unless £20,000 was immediately sent out to him, there was a serious danger that Mexican creditors would seize the whole of the company's assets. His warning fell on deaf ears, and further appeals by the directors failed to persuade any shareholders to offer more money. The resolution to wind up the company was finally confirmed at a meeting on 30 October 1848.

The sale of the company's assets took place over the next few months. Advertisements were placed in the press by the company secretary, John Phillips, inviting "proposals for the purchase of the whole of the company's rights, interests, property, and effects in Mexico."[63] The directors had already disclosed that Mexican investors had expressed an interest in acquiring the mining rights and other assets and there was also a declaration of interest by a British buyer. This was a John H. Fagan, described as "a gentleman in the City." He apparently envisaged establishing a new company to take over the Real del Monte concern. He issued a prospectus in November 1848 seeking to raise £100,000 in £1 shares. No information is available about the market reaction to this proposal but the absence of comment suggests that it received no support. Another similar attempt soon afterward to rescue the business, orchestrated by the secretary, John Phillips, also failed.[64]

While these attempts to save the company were going on in London, Buchan was in the process of negotiating with Mexican buyers. Randall gives all the details. In short, acting on behalf of the London directors in the spring and summer months of 1849, Buchan negotiated the sale and transfer of all the Real del Monte Company's rights, properties, equipment, machinery, and debts to a new association formed in Mexico and composed of prominent Mexican merchants and financiers. Buchan himself was allocated five shares in the new venture as well as being appointed its first general manager. An appraiser was hired to calculate the total value of the British company's assets. Including very large debts owed by Mexican mine owners accumulated over the previous twenty or more years, which Randall calls "semi-credits" because there was in reality no chance of them being paid, the total value amounted to $5,020,285 pesos or approximately £1,000,000 sterling. In payment, the London directors accepted 30,000 pesos or £6,000.

With obvious irony, within just a couple of years of the end of the British Real del Monte Company, the new Mexican venture had turned the business around and began making handsome profits which were magnified by "the bonanza of the century." Buchan was responsible. He had introduced major economies together with new methods and technology into the mines and he was rewarded with the discovery of "a silver-streamed bonanza that was to flow uninterrupted for many long years."[65] As a shareholder, he benefited from his success and was soon able to retire back to Britain where he enjoyed rich dividends for the rest of his life.

Finally, the winding up and liquidation of assets in London provided some meagre return for the British shareholders. While it is difficult to work out exactly the amount that was eventually paid, holders of the company's debentures and loan notes, some going back to 1827, were paid a few pounds. In 1850, for example, £4 was paid on every £50 outstanding from an 1827 loan and £2 was paid on red debentures. More payments were made the next year but of much smaller amounts, for example, 13 shillings on the £50 loan notes, and it was not until May 1857 that the final, unspecified, dividends were paid.

The Bolaños Company had been closely associated with the Real del Monte Company since its inception in 1825. It had been formed by the same group of largely Norfolk-based investors and shared much the same directors, management and to some degree, shareholders. While as we noted earlier, it had enjoyed some success in the 1830s and dividends had been paid, by the early 1840s it was making losses on its mines and found its shareholders reluctant to provide more capital. In 1844, work in the Barranco mine at Bolaños, once the Company's main area of activity, was completely abandoned. All that remained a year later, the shareholders were told, was "one Superintendent at a weekly salary of $27, and a few watchmen."[66] Attempts to sell the redundant machinery, including steam engines, found no buyers. Tools and other mining equipment were sent to Zacatecas, where the company's operations were confined to a few small mines. At some point around late 1845 or 1846, however, the directors appear to have leased a new mining property known as El Bote. Progress reports on the Bote mine received in London from the company's on-site managers appeared in the British newspapers every few weeks. In *The Times* (3 July 1848), for example, it was reported that the Bote, "as a virgin mine, promises successful results." By the following March (*Daily News*, 12 March 1849), the optimistic tone had disappeared: "The present appearance in El Bote is productive of great uneasiness, and the accounts show a heavy loss." The next few reports gave more hope, emphasizing that the erection of a steam engine and "other requisites" were progressing satisfactorily. By 27 June 1849, however, at a special meeting of the shareholders, the directors were obliged to admit that serious problems had arisen.[67] They explained that in the previous twelve months, 84,636 pesos had been spent on the Bote mine and that "this heavy outlay had been incurred in the erection of necessary works, machinery, roads etc. and the whole was in a perfect and effective state to proceed with spirit, on the production of the necessary funds." What this rather convoluted phrasing meant was that the company had run out of money and that work at the mine had been halted for lack of funds to pay the workmen. The directors, therefore, proposed that a new share issue be offered to investors, of 14,000 shares at £3 each, to raise a capital fund of £42,000, for investment in the Bote mine. A minimum sale of 5,000 shares, raising £15,000, was required for the issue to be confirmed. The shares were in the form of a loan repayable to holders from the first profits, with the shares retained as a bonus. The meeting approved the proposal. Further special meetings ratified the decision and the shares were offered in the market with £1 installments due on 31 July, 28 September and 20 November 1849.

Things did not go to plan. In a crowded meeting held on 27 September 1849, chaired by Sir Robert Price, chairman of the Bolaños Company, the shareholders were told that the minimum sale of 5,000 shares had not been met. Instead, only 3,000–4,000 had been sold, raising "about £9,000" which was significantly less than the £15,000 minimum required. The owners of the Bote mine had indicated that they could not accept much further delay and the Bolaños Company must resume work in the mine or return it to its owners. The situation, Sir Robert told the shareholders, was critical. If the money needed to develop the Bote mine was not raised, there would be no further point in maintaining the Bolaños Company. It would have to be wound up with all its properties and assets sold. In light of this situation, the directors had decided to revise the share issue, now offering 42,000 shares of £1 each, payable in three installments of 6s.8d. Several speakers from the floor voiced their opinions, some in favor and others against, and in the end the meeting voted to approve the revised share offer.[68]

It was all to no avail. The revised share offer did not attract investors and the directors of the Bolaños Company decided to dissolve the company. This decision was taken around late November or early December 1849. A final meeting of shareholders was held a few months later, on 28 June 1850, when those present were told by Sir Robert Price that the winding up of the company was proceeding. All work in Mexico had been halted. Bankers in London were owed £2,100 but it was hoped that the sale of properties and other assets in Mexico would realize enough money to pay off all debts.[69]

The largest British mining company in terms of capital and mines contracted in Mexico was the Anglo-Mexican Mining Association. Information on its activities is scant largely because its directors had fallen out with the editors of *The Mining Journal* and refused to allow their or other newspaper reporters into their meetings. Nevertheless, it is clear that it also was in trouble by the early 1840s. Most of its contracted mines had been returned to Mexican proprietors including, in 1840, the Mellado mine at Guanajuato in which the Company had invested large sums with little or no return. Not long afterward, the Mexican owners struck new veins that produced a spectacular bonanza. As *The Morning Post* (21 July 1840) put it with nice understatement, "It is to be regretted, therefore, that the works should have been transferred by the Anglo-Mexican Company to its original proprietors." No better luck was had in the next few years and by 1848, the Company was virtually insolvent. With no dividends having been paid since its formation in 1825, there was no possibility of raising additional capital and it was decided in January 1849 to begin disposing of the few assets that remained. The scale of its decline is illustrated by its finances. In the 1820s, it had raised and spent over £1,000,000. In January 1849, losses on one of its remaining mines were 102 pesos or £20. In March, losses on its Asunción mine were 26 pesos over a five-week period. Its share price by September 1849 was 5 shillings for a share on which £100 had been paid.

A general meeting of shareholders was held in London on 4 July 1849 to consider a resolution to dissolve the company and to appoint a committee to implement the winding up and sale of assets. The resolution was carried and by October of the same year, the first dividend of 10 shillings arising from the liquidation was announced. Although no information has been found on the liquidation of assets, five dividends amounting to £1 10s. were paid over the course of the next few years until March 1857, when a sixth and final dividend of an unspecified sum was paid. It was not until November 1860, however, that the liquidation committee summoned a final meeting of shareholders to tell them that the accounts were finally closed "and the assets fully administered."[70]

The fourth British mining association was the Mexican Company. Compared with the other British companies, it had always had several distinct features. It was capitalized initially at £1,000,000 divided into 10,000 £100 shares. The full capital, however, was never raised. Calls on shareholders were always modest at one or two pounds, and as a result by the late 1840s, less than £400,000 had been raised and spent. Another reason for the undercapitalization was that with every call made, a number of shareholders decided to cut their losses and not pay, thus forfeiting their shares. The result was that by 1838, only just over 5,000 shares were still valid and a third of those were held by directors.

In the first months after its formation in 1825, numerous mines had been leased in Zacatecas, Veracruz, and Oaxaca but it was soon realized that most of them were unlikely to be profitable. The directors decided to concentrate their efforts exclusively on Oaxaca, and all the mines in other areas were abandoned or returned to their former owners. For the next twenty-five years, therefore, it was in the southern state of Oaxaca where the mainly German and Italian miners recruited to work for the company tried to make profits. For the most part, they failed, and although various managers introduced new or different methods of mining and refining the ore, none proved to be successful. It was not until 1841 that the first small profit of £130 was made. That achievement was hailed as a harbinger of a future prosperity but hopes were soon dashed when the following year incurred a loss of £3,497. By 1842, the few mines that were still being worked were yielding diminishing amounts of ore and it was decided to concentrate less on mining and more on refining ore bought from local miners. The company had invested heavily in a new barrel amalgamation mill that was expected to produce an annual profit of at least £4,000 provided enough ore could be acquired for refining. For the next few years, fortunes fluctuated with some years showing a small profit and others a loss. Money was always scarce and more calls on shareholders had to be made just to meet current expenses both in London and in Oaxaca. Military conscription of Mexican miners during the war against the United States reduced the supply of ores and brought more losses, and when in 1849, the local manager at Oaxaca insisted that he had to have more funds, the shareholders decided that no more money was to be sent from Britain.

The Mexican Company survived for a few more years but shareholder patience was clearly wearing thin. In 1850, a call of 5 shillings per share was made on them but only sixty-eight shareholders with 3,155 shares between them paid. Fifty-four shareholders with 1,936 shares declined. Office and other expenses in London were reduced from £500 to £270 a year and salaries in Mexico were cut but it was all to no avail. By 1851, the balance at the company's bankers was down to £532 cash and £500 in Exchequer bills.[71] Finally, at a general meeting on 22 April 1852, the directors recommended that the company be wound up. The loss on the previous year had been £1,392 and although the manager in Oaxaca was insistent that he could still make money, the meeting accepted the directors' recommendation. A small committee was appointed to realize the sale of assets in Britain and Mexico. Those in Britain had been valued at £840 but there were debts of £430.[72] The value of Mexican rights and properties remained to be established but the Mexican manager, Pascual Fenochio, began the liquidation in July 1852. Writing to the company's agents in Mexico City, he apprised them of the situation and asked that they make a start by canceling the subscription to the daily newspaper, *El Siglo XIX*.[73] Almost nothing was left of the £400,000 capital that British shareholders had invested. The editor of *The Mining Journal* commented as follows:

> Thus do we see the extinction of another of those undertakings started during that mania to secure the precious metals of the western world which raged in all its furor from 1824–1830 which, however, instead of realizing the hopes of the enterprising adventurers, has been the grave of millions of British capital.[74]

Only one of the original mining companies was left after 1852. This was the United Mexican Company. In many respects, its fortunes had been similar to those of its counterparts. In the years between its foundation in 1824 and 1850, it had experienced highs and lows, in some years generating profits and in others losses. Like them, after at first contracting dozens of mines in several parts of the country, it gradually gave up most on realizing that they were not likely to be profitable, even with the benefits of British skills and technology.[75] The company suffered the same shortage of funds as shareholders became increasingly reluctant to pay calls or take up new share or certificate issues. After the early premiums generated during the Bubble years, the company's share price collapsed to progressively lower levels. At the height of the Bubble in February 1825, a £40 share on which only a £5 deposit had been paid, reached £160. Rapid decline followed and by December 1831, it was down to £5 for a fully paid-up share. During the rest of the 1830s, it stayed at around £4–£5 but then fell again in the 1840s until by 1851, it was at less than £3. New scrip and debentures raised some much-needed cash but the scale of the mining operations was drastically reduced to reflect the availability or lack of funds.

By 1841, the only mine still worked by the Company was the Rayas mine at Guanajuato. All other properties at Zacatecas, Oaxaca, and elsewhere had been returned to their owners and local managers were told that no more money would be sent from Britain. Investment in new mines could only come from profits made at Rayas or at the refining haciendas that the Company still owned and operated profitably. Shareholder disillusion at the paucity and rarity of dividends was expressed by Mr. Daws. £1,000,000, he noted, had been spent by the Company in Mexico "without any profit to the shareholders." It was sad to think, he continued, that "English people should have spent such immense sums of money to fructify Mexico instead of using it at home."[76]

The Company's lease on the Rayas mine expired in December 1841 and long, involved negotiations for its renewal were conducted in the early 1840s. By 1845, nothing was settled and consideration was given to a suggestion that the Company should be gradually dissolved and the assets liquidated and divided among the shareholders. The suggested winding up was not supported by a majority of the shareholders who still retained hopes of a decent return on their investments. No more money would be sent from Britain, it was emphasized, but the local manager was authorized to spend up to £5,000 from assets held in Mexico on the Rayas contract or on new mines. Eventually, terms with its Mexican owner were agreed for the Rayas mine and several new mines were acquired. Enough profit was made for a dividend of 7s. 6d. to be paid in 1849. Also hopes were always high that the claims against the Mexican government for debts, damage, and loss of profits caused by rebel incursions that we have mentioned earlier would finally be met but it was not until well into the second half of the nineteenth century that the claims were finally settled.

In the 1850s, all the mines and refining haciendas made small but steady profits that enabled more dividends to be paid and some small additional capital—£21,500—to be raised from shareholders. Over the course of time, directors, shareholders, and employees in Mexico obviously changed but the United Mexican Association survived for the rest of the nineteenth century. Changes to the structure and rules were made from time to time as company law in Britain changed. The original association formed in 1824 was registered, or re-registered in 1862. It lasted for another thirty years until it went into voluntary liquidation in 1895. Reconstructed as United Mexican Mines Ltd., it was finally closed and was removed from the Register on 29 July 1902.

NOTES

1. *Times*, September 12, 1826.

2. H. G. Ward, *Mexico in 1827*, 2 vols. (London: Henry Colburn, 1828; 2nd expanded ed., 2 vols., London: Henry Colburn, 1829), 2:107–08.

3. I. Monroy Castillo, *Sueños, tentativas y posibilidades. Extranjeros en San Luis Potosí, 1821–1845* (San Luis Potosí, Mexico: El Colegio de San Luis Potosí, 2004), 345. According to Ward, a Mr. Stokes was the resident director of the company in Catorce. He was almost certainly a relative of Charles William Stokes who arrived in Catorce around March/April 1827. Ward also says, however, that he was shown around the mine by the resident agent, Mr. MaCartney. Ward, *Mexico in 1827*, 2:484, 498.

4. D. Barclay to G. Canning, April 1, 1824, NA, F.O. 50/8, f.125. Robinson arrived in Mexico on the English ship *Tigre* on June 18, 1824 together with twenty-seven other employees of the Anglo-Mexican Mining Association; see the correspondence in SRE, 2-3-2251, fol. 5.

5. It seems that Robinson was dismissed from the Anglo-Mexican Mining Association after disagreements over management and mining matters with his colleagues in Mexico. E. James, *Remarks on the Mines, Management, Ores, Etc. of the District of Guanajuato, Belonging to the Anglo-Mexican Mining Association* (London: Effingham Wilson, 1827), 15–16.

6. The quotation is from the prospectus.

7. It is not entirely clear what happened to the Catorce Company's mines. John Kiddell, an employee of the Company for several years, is listed by Monroy as the general administrator until 1834. It may be that it took several years to liquidate all the Company's properties. See Monroy, *Sueños, tentativas y posibilidades*, 343–47.

8. NA, F.O. 50/23, fol. 145.

9. Report in the *Times*, February 7, 1827.

10. R. Pakenham to Earl of Dudley, December 24, 1827, NA, F.O. 50/36, fols. 271–74.

11. Ward, *Mexico in 1827* (1829 ed.), 2:496. Pakenham, Ward's successor as Chargé d'Affaires in Mexico, reported to the Foreign Office on October 29, 1828 that the company had ceased operations in accordance with the directors' instructions: NA, F.O. 50/44, fols. 157–58.

12. *Times*, November 10, 1828.

13. *Times*, November 28, December 2, 1828.

14. Ward, *Mexico in 1827* (1829 ed.), 2:496.

15. Cross calculated that 256 mines were contracted but by 1827 only sixty-four were being worked. H. E. Cross, "The Mining Economy of Zacatecas" (unpublished PhD diss., University of California, 1976), 181–82.

16. *Morning Post*, February 15, 1841.

17. R. W. Randall, *Real del Monte: A British Mining Venture in Mexico* (Austin: University of Texas Press, 1972), 73–74.

18. *The Era*, March 5, 1843.

19. *Morning Post*, May 2, 1845, May 7, 1847.

20. Report of meeting in the *Morning Chronicle*, February 28, 1845.

21. Cross, "The Mining Economy of Zacatecas," 212–18.

22. *Report of the Directors of the Mexican Company*, May 1, 1828 (London: J. Plummer, 1828), 8.

23. Randall, *Real del Monte*, 138–45.

24. *Times*, June 21, 1828. Letters relating to the incident are in NA, F.O. 50/46, fols. 278–85.

25. *Quarterly Mining Review* 2, no. 5 (April 1831): 184.

26. Pakenham to J. M. Bocanegra, March 5, 1829, SRE, 2-2-1905.

27. *Quarterly Mining Review* 2, no. 6 (January 1832): 346.

28. Ibid.

29. *Quarterly Mining Review* 2, no. 5 (April 1831): 139.

30. Ward, *Mexico in 1827*, 2:336.

31. *Report of the Directors of the Anglo-Mexican Mining Association Presented at the Special General Meeting of the Proprietors Held at the City of London Tavern, Monday, August 28, 1826* (London: Henry Stokes, 1826), 29.

32. A. C. Todd, *The Search for Silver: Cornish Miners in Mexico, 1824–1947* (Padstow, UK: The Lodenek Press, 1977; repr., Exeter, UK: Short Run Press, 2000), 37–38.

33. *Quarterly Mining Review* 2, no. 6 (January 1832): 346.
34. *Quarterly Mining Review* 1, no. 4 (March 1830): 488–89.
35. *Quarterly Mining Review* 2, no. 5 (April 1831): 188.
36. E. Barron to J. O'Gorman, October 11, 1833, NA, F.O. 50/80b, fols. 286–87.
37. Cited in Todd, *The Search for Silver*, 42.
38. Ibid., 93–94.
39. The file relating to Joseph Owen is in NA, F.O. 50/118, fols. 201–10.
40. Todd, *The Search for Silver*, 165.
41. R. O. Auld and J. H. Buchan, *Notice of the Silver Mines of Fresnillo* (London: Cochrane and M'Crane, 1834), 45.
42. Ibid., 18.
43. The Bancroft Library has a two-volume collection of copies of the Bolaños Company Reports (henceforth referred to as Bolaños Company Reports [Bancroft]). This quotation is from vol. 1, (1827–1833), 113.
44. Report of the Directors of the Bolaños Mining Company, June 25, 1834, Bolaños Company Reports (Bancroft), 2:5
45. Manager's Report, April 20, 1827, Bolaños Company Reports (Bancroft), 1:14.
46. *Quarterly Mining Review* 2, no. 6 (January 1832): 408.
47. Ibid., 320.
48. Randall, *Real del Monte*, 114–25, has a very useful account of the various refining methods that were tried.
49. *Diario del Gobierno*, March 7, 1839.
50. "List of Papers relative to the claims of Mr. Kinder upon the Mexican Govt.," NA, F.O. 50/111, fols. 116–17.
51. "The British and an early *pronunciamiento*," in W. Fowler, ed., *Forceful Negotiations: The Origins of the Pronunciamiento in Nineteenth-Century Mexico* (Lincoln: University of Nebraska Press, 2010), 125–42.
52. NA, F.O. 50/99, fol. 135.
53. NA, F.O. 50/52, fols. 15–17.
54. G. Ferry, *Escenas de la vida mexicana en 1825* (Mexico: Secretaría de Educación Pública, 1945), 9.
55. Letter dated Mexico, April 23, 1825 in ING Barings archive, D.E.P. 193.74.2
56. R. Pakenham to Charles Vaughan, October 10, 1827, in Sir Charles Richard Vaughan Papers, Library of All Souls College, University of Oxford, C81/2.
57. John Borrow died the same year, a victim of the cholera epidemic.
58. *Letters of George Borrow to the British and Foreign Bible Society* (London: Hodder and Stoughton, 1911). I have used the online text by the Gutenberg ebook project.
59. *Liverpool Mercury*, January 12, 1849.
60. *Times*, July 3, 1848; Randall, *Real del Monte*, 202–03.
61. *Morning Chronicle*, August 29, 1848.
62. *Standard*, September 26, 1848.
63. *Morning Chronicle*, December 6, 1848.
64. Details in Randall, *Real del Monte*, 206–09.
65. Quotations from Todd, *The Search for Silver*, 153. Buchan explains the changes he was making at Real del Monte in a long letter to William Parish Robertson, dated April 4, 1852. The letter is published in W. P. Robertson, *A Visit to Mexico* (London: Simpkin, Marshall, and Co., 1853), 234–57.
66. Annual Report of the Directors of the Bolaños Mining Company, June 27, 1845, Bolaños Company Reports (Bancroft), 2:302.
67. *Daily News*, July 2, 1849.
68. *Morning Chronicle*, September 27, 1849.
69. *Daily News*, June 29, 1850.
70. *London Gazette*, November 9, 1860.
71. Details from report of Annual General Meeting of the Mexican Company in the *Times*, June 23, 1851.
72. *Times*, April 23, 1852.

73. Manning and Mackintosh Papers, Folder 54, 27 July 1852.
74. *Mining Journal*, April 24, 1852.
75. The Company's contracted mines are listed in Ward, *Mexico in 1827*, 2:66. They were spread across Guanajuato, Guadalajara, Zacatecas, Chihuahua, Oaxaca, and the state of Mexico.
76. Report of half-yearly General Meeting in the *Mining Journal* (1841), 245.

Chapter Six

Mining Companies: Phase 2 (1830–1860)

Between 1824 and 1826, there were sound reasons to buy shares in the joint stock companies established in London to exploit Mexico's natural resources. After the early difficulties experienced with the Iturbide empire, Mexico appeared to have resolved its most pressing internal political problems with the establishment of a federal republican form of government. Presidential elections had been held without major disruption, and a popular president—Guadalupe Victoria—had been sworn into office. He had formed an administration that he at least hoped represented all strands of political opinion, thereby removing any risk of internal domestic conflict. Governments in Europe and North America were persuaded that the new nation of Mexico was stable and permanent, with no risk of reconquest by Spain. Britain and the United States had already granted full diplomatic recognition and many other nations around the world were preparing to do the same. While the economy was underdeveloped in most areas because of Spanish colonial restrictions, few observers had any doubts that the country's strategic geographical location between East and West and its apparently vast natural resources were certain to provide wealth and prosperity, not just for the Mexicans, but also for those foreign investors willing to provide the means needed to promote economic development. Investor confidence had been amply illustrated in the over-subscription of Mexico's two foreign loans floated on the London stock exchange for the nominal total of £6,400,000 or 32,000,000 pesos. Dividends on both loans were being regularly paid every quarter and there was no reason to suppose that anything would prevent them yielding a secure and handsome return to investors for many years to come.

Then, as we have indicated in the first chapter of this work, there was the highly positive image of Mexico portrayed in the diverse media of the time.

Bullock's exhibitions in his Egyptian Hall were certainly a major influence on public opinion and his own personal decision to sell his business and uproot all his family to go to live in Mexico confirms his belief in the country's and his own future prospects. His book and other publications added to Mexico's reputation as a land of opportunity, as did the many other books, pamphlets, newspaper articles, theater productions, and the myriad sources of information that the British public had at its disposal. In the light of this "Mexicomania" across the length and breadth of Great Britain, it is no surprise that those with money to invest wanted to buy shares in the new mining companies.

The favorable and optimistic image that Mexico enjoyed in Britain in the mid-1820s had changed beyond recognition a decade later. President Victoria had proved to be the first and only president for many years to come to complete his term of office. Political rivalries and conflict had become endemic in many parts of the country with military rebellion almost the accepted means of achieving office at national and provincial level. The federal constitution had been replaced with a centralist system that provoked rebellion and the separation of Texas. Liberals and conservatives fought over Church reform, education, press freedom, individual rights, land distribution, and a host of other contentious issues. Far from being a land of opportunity, the economy had made no evident progress. The national government, forced to devote almost half the budget to the army, was for all practical purposes bankrupt. In October 1827, it had defaulted on dividend payments due on the bonds sold in Britain, and Mexico's credit worthiness was thereby destroyed. As a consequence, there was almost no capital infrastructure investment financed from abroad or from within; trade was stagnating; unemployment was high; and poverty was widespread. As British merchants living in Mexico City put it in June 1836, "We have lived to witness the almost total annihilation of our trade—to see ourselves deprived of the fruit of years of toil and enterprise—to find ourselves surrounded by universal bankruptcy."[1] None of the British mining companies had prospered and their shareholders' hopes of silver bonanzas and generous dividends had all but disappeared. In sum, Mexico's image by 1836 was of a nation on the brink of anarchy and possible disintegration, liable to be absorbed completely by the wealth and expanding colossus to the north, the United States of America.

Yet this undeniably accurate picture of a politically unstable and impoverished nation is misleading as far as the British were concerned. Paradoxically, although Mexico's mining industry had proved to be a poor investment, absorbing millions of pounds of British investors' money with little or no return, the British public's faith in it, while certainly diminished, was not extinguished. Disregarding all the political and other problems facing the country, whenever there was a hint of a bonanza or talk of new discoveries of gold or silver, there were always people willing to risk their savings. Unnoticed by all those historians and others who have written of British interest in Mexico, and especially in the mining industry, there was in fact a second wave of investment in the 1830s and 1840s when several new companies were set up in London to develop or invest in Mexican mines. While not on the scale of the company formation in the 1820s, either in terms of manpower or finance, the new companies were fully incorporated with their shares offered for sale on the London stock exchange. Moreover, notwithstanding the negative image of Mexico in the 1830s, there was no shortage of buyers for the new shares. The British public continued to believe in Mexico's economic potential, and rather like the Mexicans themselves, they had concluded that once the domestic political issues had been resolved, prosperity and profit were bound to follow. This second phase of company formation began with the Peñoles Gold Mining Association.

PEÑOLES GOLD MINING ASSOCIATION (1831–1837)

This Company's original promoter seems to have been a German mining engineer, Frederick Anton Weichsel. He had worked in Mexico for several years for some of the other British companies and during his time in Oaxaca, working for the Mexican Company, he had noted the existence of gold deposits in the small village of Santa María Peñoles, a few miles north of the city of Oaxaca. These had been worked by the Spaniards but had long since been abandoned. Weichsel appears to have acquired some of these from a local miner, Juan Hernández, for the sum of 5,700 pesos and then decided to seek British investors to finance renewed mining operations.[2] In 1830, he traveled to London where he approached, among others, Henry English, editor of the very influential *Quarterly Mining Journal,* and George Walters, a London businessman. Weichsel provided them with ore samples, claiming that his own tests had indicated a yield of three *adarmes* (one sixth of an ounce) of gold per load of ore.[3] Other assays done in London suggested that this was a considerable underestimate and that as many as twenty to thirty *adarmes* might be obtained. He also estimated that he needed no more than £5,000 to bring the mines into production and that returns of at least £11,000 per annum could be expected. Although details are scarce, it seems that English and Walters were attracted by his proposition and having signed a contract with Weichsel in December 1830, they began to recruit investors. By early 1831, a Board of Directors had been recruited and an office address rented at 5 Baldwin's Court, College Hill, London. The first directors included John George Druke, Lloyd S. Baxendale, S. Chambers, Henry English, J. T. Jeune, and J. Milroy. Others were added later. The prospectus was published and English provided a summary in his *Quarterly Mining Review.*

The capital of the new company was to be £12,240 in 1,020 shares of £12 each. The first installment was £4 and payable to the bankers, Sir J. W. Lubbock and Co. Subsequent installments were to be no more than £2. There were to be the usual half-yearly meetings of shareholders in May and November.[4] Weichsel, described as an "experienced and intelligent miner" was appointed manager of the new venture.

Weichsel left Britain for Oaxaca toward the end of 1830. With financial backing secure—by November 1831, £6,457 14s. had been "remitted to Peñoles for Administration, including purchase of the Mines and Contingencies"—he was able to acquire several mines for the Company.[5] Three of these—Guadalupe, Santa Barbara, and San Luis Gonzaga—were formally conveyed by Weichsel to the Company on 19 September 1831.[6] Other mines in the Peñoles locality were soon obtained including "a gold mine called Our Lady of La Soledad in the village of Santiago Tlazoyaltepec" which was acquired from Nepomuceno Ramírez for just 300 pesos.[7] Machinery and supplies ordered in Britain—quicksilver, iron, copper, stamp heads, tools, and sundries—were shipped and while awaiting their arrival, Weichsel began the construction of a stamping mill, amalgamation plant, and "a dwelling house," presumably for his own use. He also recruited others to help him in the mining operations. Among these were Robert Henry, John Sadler, George Quin, and J. Sheudhart. Finally, there was Edward Muhlenpfordt, a German mining engineer who had been recruited in Hanover in 1826 to work for the Mexican Company. He took advantage of his time in Oaxaca to study and make drawings of the now famous Mitla archaeological site, about which he wrote a work entitled "The Palaces of the ancient Zapotecos at Mitla, and some other antiquities found in the state of Oaxaca." Years later, he also published a book about his time in Mexico.[8]

With funds, equipment, and personnel in place, Weichsel began the search for gold and over the next few years, he and his colleagues sent back regular—at times weekly—reports on their progress to the directors in London. Extracts and summaries of these letters, which included the weekly *rayas* or financial accounts, were published by English in his *Quarterly Mining Review* and the full text was included in the annual reports prepared for the meetings of shareholders.[9] Weichsel, Henry, and Sadler were the main correspondents in the early years and were always ready to reassure their investors back in Britain that "success cannot be doubted." There were, Weichsel wrote on 2 June 1831, "gold veins running in every direction where all, or nearly all, show an alloy of gold of two *adarmes* and upward, even on the surface."[10] On the other hand, it was also soon evident that problems were being encountered. The rainy season was prolonged and unusually severe in 1831 causing difficulties in the transportation of essential supplies, especially timber that had to be brought from mountainous areas some distance from Peñoles. When heavy rainfall did occur, flooding in the mines was always a serious problem. In contrast, the following year, 1832, was unusually dry with almost no rain after the middle of July. The drought had two effects. First, there was insufficient water for the mill and secondly, the maize crop was reduced which brought increased prices for animal feed as well as food production. 20,000 adobe bricks had to be manufactured for the various buildings and labor was always short and expensive. With little or no tradition of mining in the area, it was found to be difficult to persuade the largely Indian population to leave their fields to work in the mines and wage rates had to be increased to attract them. By April 1831, Weichsel reported that he had "only one hundred people at work." Unforeseen events compounded the problem. A serious fire at the nearby village of Santiago destroyed the local church and forty houses with the result that "Indian working people are scarce." Machinery broke and caused delays:

> Nearly the whole of last week was lost by the spindle of the cog wheel having broken, which, with other repairs and improvements that were found necessary, added to the feasts of Peñoles, it may be considered that nothing was done in the way of making gold.

Grinding of up to forty loads of ore a day caused heavy wear on the machines in which "the heads and plates will soon want to be replaced." Machinery for the stamping mill and amalgamation machines en route from London were delayed. Another essential for the mining operation was mules. These had to be bought in large numbers because it was found that none were for hire. On the more technical side of the work, iron and copper pyrites were found in some of the ore and this caused problems in the amalgamation process with "a considerable loss in the quicksilver." On many occasions, the ore proved to be of an unexpectedly poor quality. Writing on 6 October 1832, Weichsel was clearly disappointed with the results of his work:

> I am grieved to say, I have been sadly disappointed, as we got no more than 23 oz. 12dwt. 12gr.troy, from not less than 400 *cargas* of Guadalupe ores, part of which was of so good a quality, as to let me hope, that our produce would be not less than 50oz. I cannot account for this disproportion between my expectation and the produce of the last "Molienda."[11]

Another constant problem or complaint regularly expressed by Weichsel and his managers was lack of money. Funding for the whole mining venture—salaries and wages, accommodation, food and other supplies, animals, and a range of miscellaneous expenses—was channeled from the company's bankers in London, Sir J. W. Lubbock and Co., through British mercantile companies operating in Mexico City and Veracruz. The two most used were Montgomery, Nicod, and Co., and Manning, Marshall, and Co., which had offices in the capital and at the port. In October 1832, for example, Sadler reported that "three cakes of gold," together with a recently produced one from the Guadalupe mine, were being left with Montgomery, Nicod, and Co. for dispatch to London. Having no funds left at Peñoles, however, they had already valued the gold at 1,000 dollars which they had drawn on Manning and Marshall. Just two weeks later a further "thirty ounces troy of pure gold" were ready for shipment. Money, however, was always scarce and at times workers had to be laid off because they could not be paid. Opportunities for acquiring new mines or for major works such as drainage adits in the existing mines were missed or postponed and as Sadler put it, "it is a pity that the capital necessary should have been stated in the first instance at so low a figure: if it had been made £15,000 instead of £10,000, the disappointment, if any at all, would have been less."[12]

The accounts included in the report presented to shareholders on 6 May 1833 reveal that the 1,020 shares issued at £12 a share had all been taken up, yielding £12,240. Gold received from Peñoles was worth £372 16s. 7d. but bills had been drawn in Mexico against this for £350, leaving a balance of just £22 16s.7d. With further small sums from interest, the total capital raised for the Company was £12,305 13s.6d. Of this, £10,315 had been sent to Peñoles to pay for the mines and other contingent costs. Travel expenses for Weichsel, Sadler, and other employees had cost £468. Office expenses in London, including salaries, rent, postage, furniture, and incidentals had absorbed another £567. Quicksilver, machinery, stamp-heads, tools, iron, and copper with freight and insurance added took £670. All that was left in hand was £281.

Under pressure from the directors for better results and well aware that the first two years of operations had produced only "trifling amounts of gold," Weichsel looked for other mining properties. In January 1833, after what seems to have been protracted negotiations, he persuaded a local mine owner, José Pedro Contreras, to sell one of his mines to the Company for 1,000 pesos. This was located about twelve miles from Peñoles in the village of Santiago Tlazoyaltepec, where the company already had one mine. The new property, known as Nuestra Señora de la Angustias, was contracted on 19 January 1833. Unusually, the notarial document conveying it to Weichsel gives precise details of its location. It was "on lands at Tlazoyaltepec in the village of Santa María Peñoles about half a league from the town on the south side of the road heading for the capital and in the ravine known in the Mixtec language as Ynosite."[13]

Weichsel had high hopes for the mine but before they could be realized, he died in the summer of 1833, almost certainly a victim of the cholera epidemic that swept across Mexico for much of that year. It is interesting to note that he left his estate to his Mexican servant, Juan Miguel.[14] The directors appointed John Sadler as the new general manager and his main assistant seems to have been George Quin, and both of them now began to send the regular reports to London. Money was needed to develop the Angustias mine and the directors decided on a new share issue. 3,060 shares of £5 per share were placed on sale, with a deposit of £1. Henry English, who chaired the half-yearly meeting of shareholders on 3 November 1834 told them that the whole issue had been subscribed and that £1,000 had already been made available to Sadler. Work continued or was even accelerated in the mines but the yields continued to be disappointing. The shareholders' meeting of 4 May 1835 was told that 5,858 *cargas* of ore had been ground in 1834 yielding "a standard of 1⅞ *adarmes*." The profit over recent months was about £150 a month. At the next meeting, and perhaps reflecting growing disillusionment, Henry English noted in his new publication, *The Mining Journal* (7 November 1835, p. 87) that "the report of the directors, although not the most cheering, gave evident satisfaction to the proprietors present."

In fact, dissatisfaction was growing among the shareholders. Although another mine, the San Juan, was contracted by Sadler for the company in 1836, it is clear that the shareholders were unwilling to invest any more money. The price of the original shares, fully paid up at £12, had generally held up in the market throughout the years from 1831 to 1835 reaching £13 10s. in August 1833 but the new share issue did not fare so well. In fact, although the £1 deposit was paid, only one further call of 10 shillings was successful. The second share issue, therefore, raised only £4,590 (3,060 shares at £1 10s. each) instead of what had been hoped would be £15,300 (3,060 x £5). Threats of forfeiture of shares on which calls were not paid had little effect and in April 1837, the directors announced a meeting to consider dissolving the company.[15]

No record of the final meeting has survived but with no further mention of the Peñoles Gold Mining Association in the press or elsewhere, it seems certain that the company was wound up in the summer of 1837. While in comparison to the huge sums of money raised by those British companies formed in the 1820s, the amount lost by investors in the Peñoles venture was not large at around £17,000. No information on the identity of the individual shareholders has been found but it is likely that the largest investors were the original directors such as Henry English. As far as we know, he and his fellow directors/shareholders received no dividends and certainly can have made no money trading in the shares. As for the workforce in Oaxaca, Sadler and Quin both remained in the city. Sadler set up a mercantile company but was still involved in mining in the 1840s. In February 1843, he bought twelve shares in a silver mine near the village of Ixtepeji. A few months later, he acquired a refining hacienda and he was still investing in mines in 1846.[16] George Quin, an Irishman, also became a merchant and was in business in Oaxaca until at least 1848. They became part of a small group of British residents—James Innes, Edward Turnbull, James Bickett Hampshire, John Smith, and Frederick Montgomery among them—who together with the German miners such as J. Sheudhart, Charles Gustave Kaufmann, and Adolfo Hegewisch had hoped in vain to make their fortunes at Santa María Peñoles de Oaxaca.[17]

THE MEXICAN AND SOUTH AMERICAN COMPANY (1835–1859)

We now turn to another British company formed in London in April 1835. This was the Mexican and South American Company. Its aims were summarized in the prospectus as follows:

> The entering into contracts with the mine proprietors to supply them with advances of money and stores, on the principle of what is termed in Mexico "Avío de Plata." By these engagements the Contractors anticipate to the mine owners, from time to time, the forthcoming produce of their mines, receiving in consideration thereof the gold and silver produced thereby, at certain fixed rates, so regulated as to yield a handsome return for the money advanced.

The company, in other words, was to provide finance for new mining ventures in Mexico rather than acquiring and operating mines on its own account. The return on capital, it suggested, would be of the order of 1½ to 2 percent a month, a rate that "is not unusual remuneration for the use of capital in Mexico." In addition to financing of new mines, the company also contemplated "dealing in bullion, and generally in metals, and stores for mining purposes." The capital of the company was to be £100,000 in 10,000 shares of £10 each, payable as usual in installments. An annual dividend of 6 percent was guaranteed or "at once fixed." Finally, the founding directors were John Schneider, Henry Ewbank, and J. D. Powles, all three of whom were already prominently involved in other Mexican mining companies, notably the Anglo-Mexican Mining Company and the Anglo-Mexican Mint Company.

If there were specific reasons or opportunities known to the founders of the company, we do not know what they were but it was very quickly in business and over the next few years it invested in several new Mexican ventures. The first of these was the Zacatecas Mining Company.

ZACATECAS MINING COMPANY (1836–1847)

Little is known about this venture, but the survival of one of its share certificates, dated 1 August 1836, provides some useful information. It has the text of the prospectus, which opens with these words:

> The Mexican and South American Company has had proposals made to it, to enter into a Mining undertaking in Zacatecas, in the United States of Mexico, of a peculiarly favourable character. The district in which it is situated is the same as that in which the Mines of Fresnillo are situated; and the description given of the Property, by those who have examined it, is in no degree less favourable than that formerly given by the same parties of the Fresnillo Mines, the productiveness of which has far exceeded the expectations formed of them.[18]

It was not the intention to work the property but rather to transfer "the engagement to other parties, looking only for such compensation in the transaction, as in a mercantile point of view it is entitled to." As the directors of the Mexican and South American Company were also on the Board of the Anglo-Mexican Mining Company, the property would in the first instance have been offered to the latter, but its finances did not permit it to accept the offer. But, the company would take an interest appropriate to its financial condition. The capital of the company was to be £70,000 in 14,000 £5 shares. Four thousand shares were to be appropriated to the Anglo-Mexican Mining Company, 2,000 to the Mexican and South American Company for special purposes connected with the acquisition of the property, leaving 8,000 available for public subscription. Finally, the company's bankers were to be Barclay, Bevan, Tritton, and Co. and it was to be managed by the directors of the Anglo-Mexican Mining Company, namely John Schneider, Chairman, Matthew Harrison, Edward Hurry, Benjamin Shaw, Henry Ewbank, Charles Herring, J. D. Powles, and Alderman Thompson, M.P. Prominent figures in London's mercantile community, we have encountered all of these names several times before in connection with the mining companies.

The origins of the Zacatecas Mining Company may have been connected to recent events in the state of Zacatecas in northern Mexico where, as we have seen already, the British mining companies had extensive operations. Francisco García Salinas, a liberal and strong federalist, had been elected governor of the state in 1828. Anxious to protect his territory from incursions by Indian tribes from the north, and his state's sovereignty from centralist politicians in Mexico City and elsewhere, he recruited and equipped a strong force of local militia. It was quickly evident to him, however, that he lacked the revenues to maintain and strengthen his militia or to carry out the many public works that he planned. Hence, he decided to exploit the silver mining industry on the state's account rather than allow all the profits to pass into the hands of private individuals or foreign companies. He organized the development of mines in the Fresnillo area, notably around the Proaño hill that had produced very substantial revenues in the colonial period. Abandoned years earlier and with the shafts flooded, a major program of remedial work was undertaken and within a year, production at the mines began. At first, all went well but by 1833, many problems had arisen. Political conflict and virtual civil war for much of 1832 had disrupted mining operations and the cholera epidemic in 1833 devastated the workforce at the mines. Faced with such difficulties, García decided to seek outside help and in February 1833, his legislature authorized him to appoint commissioners to go to London to look for investors. The two men he appointed—Robert Ogilvie Auld and John Hitchcock Buchan—were already in Mexico, where they had been working for other British companies. They were significant figures in the British interest in Mexico and more will be said of them later. For the time being it suffices to note that they were in London early in 1834 and promptly published a long and persuasive prospectus that described in detail the opportunities offered in the Proaño mining district, where they said a profit of 32 percent on invested capital was easily attainable.[19] They outlined the terms of the contract on offer. These, in sum, were that the Zacatecas government would cede possession of the mines for a term of twelve years; guarantee protection of the mine operators and their work force; ensure that there would be no increase in taxes on silver; and no forced loans would be levied. Finally, profits were to be divided equally between the government and the contractor.

It seems probable that these proposals and promise of high profits would have attracted considerable interest in London but events in Mexico promptly stalled any agreements. García's fears for his state's autonomy were fully justified in late 1834 and 1835 when centralist republicans gained control of the national government and began the process of dismantling the federal structure in favor of a centralized republic. García tried to resist but on 11 May 1835, he and his militia were soundly defeated in what became known as the battle of Guadalupe. The victorious military were led by General Antonio López de Santa Anna, then, and several times later, the President of Mexico. He allowed his men to ransack the city of Zacatecas, including looting from several of the British mining companies, and he himself took possession of the Fresnillo mines. He sold the existing stocks of silver and ore worth about 1,000,000 dollars to his friends. Faced with public outrage at his obvious venality, he announced in September 1835 that there was to be a new company to work the Fresnillo mines, called the Compañía de Minas Zacatecano-Mexicana. Half the twenty-four *barras* or shares were reserved for the government with the other half available to investors. Each of the public *barras* was divided into ten shares of 10,000 pesos each. In short, there were 120 shares of 10,000 pesos each available to investors.

We do not know if the Mexican and South American Company invested in this new venture. The lists of shareholders published with some of the first annual reports do not mention any direct British interest but it is always possible that nominee Mexican shareholders were acting on behalf of British investors. Shares were also bought and sold and by 1848, there was at least one British shareholder. This was William Phelps, a merchant who had been in Mexico since 1827. He received his dividends from Manning and Mackintosh, who were the company's agents in Mexico City.[20] Also, it is clear that numerous Britons were employed by the company. The Mexican passport and security pass records indicate that there were at least twelve in 1837, including John Penberthy, described as an engineer, and William Pollard, the chief accountant.[21] Nine years later, fifteen names were included in a list of Englishmen working "on machinery at the Proaño mines"[22] In 1838, Richard Rickard brought a steam engine and seven engineers from Britain to operate it. Mining tools costing 2,672 pesos were imported from British suppliers, and regular shipments of mercury were ordered from Britain via the British mercantile house based in Mexico City, Cotesworth, Cochrane, and Company.[23] Hence, while there may have been no British capital invested directly in the Mexican company, there was considerable British participation. Other opportunities for investment in the Zacatecas region were also evident and we can assume that the founders of the Mexican and South American Company not only read the report by Buchan and Auld but also met the two men while they were in London. They told them of abandoned mines in the Plateros district, just a couple of miles from Fresnillo. It was several of those mines that were contracted for the new Zacatecas Mining Company that coexisted with its Mexican counterpart, the Fresnillo-based Compañía de Minas Zacatecano-Mexicana.

Very few of the records of the Zacatecas Mining Company have survived, for example, none of the company reports that were presented to the annual meetings of shareholders. Fortunately, however, Henry English continued to send his reporters to the company meetings, and details of its operations were published in his *Mining Journal*. The person chosen by the Mexican and Spanish American Company directors to travel to Mexico as the mine manager was John Henry Bodmer. He appears in the Mexican passport records on 24 November 1836, described as "Commissioner for the English Mining Company called the English and South American Company." He was accompanied by Edward Bodmer, probably his brother, who also appears in the passport records on the same date as "resident at Zacatecas for the English and South American Company." Other experienced miners, especially Germans already working for mining companies in Mexico, were recruited. They included Christian and Emil Schleiden, and F. Schuchardt. Although we have no details of the contracts, several mines in the Plateros district were acquired from their Mexican owners, including the Leona, San Cristo, Prosperidad, and Santa Teresa Trinidad. Work on behalf of the company also seems to have been undertaken in the Descubridora shaft of the Valenciana mine.

The first accounts of mining operations sent back to London seem to have been sent by Schmann and Penny, the company's agents in the city of Zacatecas.[24] They reported that Bodmer had found that he had insufficient funds to start work. Mr. Schmann had decided, therefore, to inspect the mines for himself and was so impressed with their potential that he had advanced £5,000 to Bodmer. Work began soon afterward and in May 1838, Bodmer told his directors in London that early results in all the mines had given him great hopes of "imminent success." He had already been obliged to buy another 100 horses and was actively seeking to buy more and the fodder to feed them.[25] Over the next year, he and Schuchardt sent encouraging letters back to London, but it seems that the directors were quickly disappointed that no silver shipments had been made. Possibly because they had run out of money and were unable to raise any more by selling shares in the market, they decided at the beginning of 1839 to suspend operations and to send out a special commission to evaluate the potential of the mines. Reporting on the situation in *The Mining Journal* on 22 February 1840, English noted that the commission "have reported very favourably" on the mines and on the company's prospects. All that remained was for the shareholders to decide if work was to be recommenced. That was a matter of some urgency because since the suspension of work a year earlier, a large sum of money had been and was still being spent on the *amparo*. This was a legal process by which the right to retain mines was sustained by keeping a number of men at work on the properties.

Special meetings of the shareholders were held in March and August 1840 and according to *The Mining Journal* they decided, presumably on the strength of the commission's report, to reopen work on the Cata de Plata mine.[26] It had been mined for some time in 1837 and 1838 by the Mexican Company but had been abandoned since then. Water had flooded the shafts but it was now thought that an appropriate steam engine would easily solve the problem. Large profits, English added in his notice, could now be expected.

English's optimism was not justified because it appears that the Zacatecas Mining Company directors decided not to recommence work in the Mexican mines. There is no further reference to the company until a notice in *The Times* on 19 February 1847. This announced that a final dividend of four shillings per share was to be paid on the shares on and after Friday, 25th inst. This was probably paid from the proceeds of winding up the business. It seems clear that the investors had lost most of their money.[27] Christopher Richardson may have been one of those investors. In a pamphlet published in 1855 in which he bitterly attacked the activities of John Diston Powles and his family, accusing them of all manner of nefarious and corrupt conduct in their promotion of companies, he stated that the Zacatecas Mining Company had never paid a dividend and that when it was wound up, shareholders had received only nineteen shillings for every £5 10s. subscribed. He also states that the company was wound up in 1841, and given the lack of any further reference to it after that year, this does seem probable.[28] It may also be relevant, as we will see later, that the Mexican and South American Company directors decided to liquidate all their Mexican investments in 1847.

Little is known of the subsequent careers of the company's employees. As there is no further mention of him in the Mexican passport and security pass files, it is likely that John Henry Bodmer returned to Britain.[29] His brother Edward, however, stayed in Mexico and over the next few years was in Durango and Guanajuato, working for other mining companies. By the 1850s, he was the British vice-consul in Taxco, where he also had shares in silver mines. In January 1861, while apparently trying to rescue a prisoner being tortured by rebel troops, he was shot and killed. He left a widow and five children. F. Schuchardt had resigned at the end of 1838 and returned to Europe.

GUADALUPE Y CALVO MINING COMPANY

The next venture in which the Mexican and South American Company decided to invest its shareholders' money had more romantic origins. It is a good story and by no means unique in the annals of Mexico's mining industry, which has had several accidental finds that led to bonanzas. Briefly, the events are said to be as follows. In October 1835, an Indian muleteer by the name of Miguel Urías was taking his animals along the plain of El Zorrillo, high up in the mountains of southern Chihuahua. One of the mules went astray and Urías looked around for a stone to which he could tether the others while he went to find the stray. He saw what he thought was a suitable rock but when he moved it, he noticed that it was much heavier than he had expected. Then, using an old trick taught him by local miners, he licked it. He knew immediately that it contained minerals and a quick survey of the surrounding ground convinced him that he had come across a mineral-rich area. On 30 0ctober 1835, he went to his nearest local authority, the *juez* or judge at Nabugami, and staked a claim, calling it *Nuestra Señora del Rosario*. Word of his find soon leaked and local miners rushed to register their own claims adjacent to that of Urías. Soon there were forty-two claims registered, each giving rights to mining over an area of around ten square meters.[30]

The news quickly spread far and wide and people arrived in such numbers that within months the population exceeded two thousand. Streets were laid out, houses in the form of log cabins were built, and supply lines for food and other necessities were quickly established with the few neighboring communities in the vicinity. Then, in December, the *alcalde* or mayor of the nearby village of Dolores arrived. He insisted that the mining area discovered by Urías was in his jurisdiction and that all claims had to be registered with him.

Not wanting to be involved in a jurisdictional dispute between rival authorities, Urías and the other miners decided to request the creation of a new municipality. For it, they chose the name of Guadalupe, patron Saint of Mexico, and Calvo, after the then governor of the state of Chihuahua, General José Joaquín Calvo. Their proposal was quickly accepted by the governor and the town of Guadalupe y Calvo was founded.

Guadalupe y Calvo is 2,345 meters above sea level, about 180 miles southwest of Chihuahua City, and near the borders of Durango, Sinaloa, and Sonora. With a bitterly cold climate in winter and oppressive heat in summer, it is even today a remote and still predominantly Indian community. In 1835 it was accessible only by mule or horse along narrow and often precipitous mountain tracks, but the lure of gold and silver was always a powerful incentive to overcome such difficulties. As Ward put it in his book, "these wild and distant districts are visited only when the fame of their riches attracts adventurers from the towns."[31] By 1836, more and more "adventurers" had arrived and the population rose to over 4,000. Among the newcomers in April 1836 were four Britons: John Hitchcock Buchan, Robert Ogilvie Auld, James Randell, and Henry Alexander Mackintosh. All were experienced miners who had worked for the British companies in Real del Monte, Guanajuato, and Zacatecas. They had traveled several hundred miles from Zacatecas on horseback over rough and dangerous mountain trails, risking ambush and attack by marauding Comanche and other Indians, to find out what was going on at Guadalupe y Calvo. We have briefly met Buchan and Auld before when they appeared in London in 1834 as commissioners of the Zacatecas governor, Francisco García, who had sent them to Britain to find investors for his state's mining industry. Some biographical details are now useful. Buchan was born in London in 1803, the son of a Captain in the Royal Navy. He graduated from the Royal Military College at Woolwich but while awaiting a post in the British army, he was offered the chance the join the 1825 Colquhoun-led expedition that was charged with taking some 1,500 tons of equipment to the Real del Monte mines. The story of the heroic and sometimes tragic experiences of the expedition with all the immense difficulties faced after its arrival at Veracruz has been told elsewhere and we do not need to repeat it here. Buchan contracted yellow fever but was one of the few who survived it. He spent the next few years working for the Real del Monte Company, learning about mining for silver from the Cornish miners who had accompanied him. By 1836, he was an experienced and trusted miner, well known in Mexico but also just as importantly, in the financial and mercantile communities of London.

Buchan's companion and close friend, Robert Ogilvie Auld, had also been in Mexico for several years. Born in Glasgow in 1806 to a leading Scottish family, he obtained a post as personal assistant to Captain James Vetch, one of the commissioners of the Real del Monte Company, and he sailed with him from Liverpool in March 1824, together with the thirteen others who made up the company's first party sent to Mexico. Working in an administrative capacity at Real del Monte, Auld quickly learned the mining business, becoming especially well informed in the refining and smelting processes. As we have seen, he was in London with Buchan in 1834 and while there he took the opportunity to marry Mary Thomas on 10 September. He then returned to Mexico. Two of his brothers, Stuart Alexander and William, had also arrived in the country in 1832, and a third brother, Thomas Reid, would arrive a few years later.

The third Briton who made the dangerous journey across the Chihuahua mountains was James Randell. It is not known when he arrived in Mexico but he was working for the United Mexican Mining Company in Guanajuato and Zacatecas from at least 1830 to 1834. More is known about the fourth member of the British party, Henry Alexander Mackintosh. He had arrived in Mexico around 1824, probably to work for the Real del Monte Company. By 1831, he was the personal secretary of the then senior manager at Real del Monte, Charles Tindal. Few details of his work or personal life are available, but we know that he was involved in one notorious incident in 1831. He was put in charge of an escort to protect a silver convoy dispatched from Real del Monte to the Mint in Mexico City. The journey to the capital with the cargo of silver bars went without incident but on the return journey, with the mules loaded with silver coins, he and his men were attacked by bandits. Several of his men were killed and Mackintosh was himself badly wounded in the fighting. He survived his injuries and was later given a £200 bonus by the London directors of the company as a reward for his gallantry.[32]

The Mackintosh family will appear often in future pages of this work, and it is appropriate at this point to explain the background of their involvement in Mexican affairs. In 1824, Robert Manning and William Marshall arrived in Mexico as agents of the British firm of Barclay, Herring, Richardson and Co., a partnership involved in trade but also in finance, notably winning the contract to administer Mexico's second foreign loan in 1825. Manning and Marshall soon set up their own partnership in Mexico City, with a branch in Veracruz, and quickly became among the most important companies operating in Mexico. While always continuing their own business as import/export merchants with connections in London, Paris, Madrid, and New York, they were soon acting as agent, representative, intermediary, banker, treasurer, and post office for British merchants throughout the country. In April 1833, probably on the first stage of a return to Britain, Manning set sail from Tampico, on a ship named *Mexico*, heading for New Orleans. The ship's crew mutinied, and killed the captain and all the passengers, including "the eminent merchant, Mr. Manning, of the house of Manning and Marshall."[33] Following this incident, the name of Ewen Clark Mackintosh begins to appear regularly in connection with Manning and Marshall. Born in London in

1802, Ewen was the older brother of Henry Alexander.[34] He first appears in the Mexican passport records in 1827 when he sought an internal passport allowing him to travel freely around the country. It is not clear, but he may have already obtained a clerical post with Manning and Marshall, and after Manning's death, he was promoted in 1834 to chief clerk. His brother, Henry, arrived in the capital at about the same time, having obtained the post of financial manger of the city's poorhouse.[35] Manning and Marshall continued to operate under that name until 1842 when Marshall decided to retire and return to Britain, leaving Ewen as the senior partner in the company, henceforth known as Manning and Mackintosh.[36] By then, he had married a Mexican, Teresa Villanueva, and in 1839 he had been appointed British consul in Mexico City. Undoubtedly a very energetic and skilled merchant, Ewen was involved in a myriad of business activities, becoming a major and controversial figure in Mexico's industrial, commercial, and financial circles. There were two other Mackintoshes. Thomas, a younger brother, first appears at Guadalupe y Calvo in 1836. Finally, there was William, who also worked at Guadalupe y Calvo and other places where Ewen had commercial interests. He was Ewen's nephew, but which of the brothers was his father is unknown.[37]

According to the leading authority on the history of Guadalupe y Calvo, Francisco Almada, the British party arrived in the new town in April 1836. The local miners, including Urías, had already found that they lacked the financial resources and possibly expertise to work their claims profitably. Buchan and his compatriots, having inspected and evaluated the claims, were clearly impressed with what they found and at once set about acquiring the mining rights that the local miners were happy to lease to them. On 25 and 27 April 1836, contracts were signed with the local miners represented by Trinidad Meléndez, whereby the rights to a vein two hundred yards long were transferred to Buchan and Auld for a term of seven years. Meléndez was assigned a one-eighth share of future profits, and 23,500 pesos were spent in acquiring the mines and a refining hacienda.

To finance, manage, and develop their property, Buchan and his companions needed to find investors who would put up the large amount of money that was required. A journey to Mexico City was essential but it was almost 1,000 miles away. We do not know when and how he made the journey, but just a few weeks after arriving in Guadalupe y Calvo, Buchan was in the capital. There he issued a prospectus on 19 May 1836 announcing the formation of the Guadalupe y Calvo Mining Company.[38] He described in some detail the veins that had been rented from their local owners and the current yield from them, emphasizing that it was confidently expected that they would be "the richest so far discovered." The capital of the new company was 300,000 pesos, or £60,000 approximately, divided into thirty shares of 10,000 pesos (£2,000) each. The first installment was 2,500 pesos (£500) and there were to be three further installments of the same amount payable at intervals of one, three, and six months. Buchan was the general director of the company; Robert Auld, the accountant; Henry Mackintosh, administrator of the refining plant; Manning and Marshall, agents and treasurer; and finally, the company's agent in Durango was German Stahlknecht. A short time later, Buchan and Auld acquired the rights to another nearby vein and for this they formed the Mining Company of El Zorrillo. Its capital was 500,000 pesos (£100,000).[39]

Starting two days later in Mexico City on 21 May 1836, a series of legal documents setting up the new company were drawn up by the public notary, Ramón de la Cueva. Although it is difficult to interpret the notarial records, it is clear that the main promoters of the company were Buchan, Auld, Mackintosh, and James Nelson Shoolbred. Two companies—Manning and Marshall and Byrns Hooten and Co.—were also involved. Manning and Marshall agreed to act as treasurer and agent for the new company, finding shareholders and reimbursing Buchan and Auld the 23,500 pesos they had already spent. Partners and shareholders were easily found from leading figures in the metropolitan financial and entrepreneurial élite, for example, Manuel Escandón and his business partner, Juan Manuel Lasquetty; Pedro Ansoátegui and Gregorio José Martínez, both partners in the firm of Martínez del Río Hermanos; Lorenzo Carrera, a Spanish financier; José Delmotte, a Mexico City merchant of unknown origin; Francisco Morphy, mine owner and investor in many other enterprises; and members of the Morán and Moreno families, both at the pinnacle of Mexican society. Finally, there was Edward Pine Coffin, probably a United States citizen who is described in the passport records for 1835 as "commissary in the service of Great Britain."

Ewen Mackintosh had assembled a powerful and wealthy group of investors and although we do not know which of the original partners remained shareholders—mining shares were often traded—the company was fully funded. Buchan was confirmed as general manager and Mackintosh as director of the new refining plants that were to be built. Auld became director of the El Zorrillo Company, but there is no information available about its formation or shareholders. Silver, and to a lesser extent, gold production was already under way and by June 1836, 403,000 pesos worth had been mined. With the additional investment in machinery and men, output rapidly increased and over the next few years, more than 6,000,000 pesos worth of gold and silver was produced.[40] The wages of the miners amounted to 15,000 to 20,000 pesos a week in 1837 and the prosperity brought by the mines was quickly reflected in the town of Guadalupe y Calvo. A church, a cemetery, a school, a prison, an assay office, and municipal offices were built. The population continued to rise and according to the 1837 census, it had increased to 7,666. Among the newcomers, there was a growing number of Britons. In

1838, for example, a list in the passport records has the names of nineteen British citizens applying for security passes. Another arrival was John Buchan's brother, Stephen Douglas Buchan. Robert Auld's brother, Stewart Jolly Auld, also arrived at about the same time. Many more people appeared in following years. Some, but not all, were recruited from other mining companies, for example, William McLorg, employed at Real del Monte and Bolaños since 1827, and Daniel Williams, who had worked for the United Mexican Company at Oaxaca and elsewhere since 1832. Richard Ninnes, Richard Cendon, Joseph Bray, and Joseph Scobie, all described as *mecánicos*, were granted their permits on 16 February 1837. Other professions also arrived. Almada tells us of James Cheine [sic.] or Cheyne, Joseph Salding and John Jemison, who were the first doctors to practice in the town. Jemison conducted experiments with smallpox vaccine, successfully vaccinating some 600 people just prior to an outbreak of the disease.[41] Joseph Davies was a British carpenter who constructed a bridge across the river and donated it to the town. Charles Bowring spent at least some of his time making astrological and geological measurements, and the results of his labors can now be consulted in the Royal Geographical Society.

Finally, there was the American Gilbert Pell. He arrived at the beginning of August 1837 in search of employment. Very unusually, some of his personal correspondence has survived and it provides us with much interesting detail of life in the new mining community.[42] The climate bothered him. Writing to his brother-in-law, who was at Zacatecas, he told him, "it has rained here every day since my arrival, and so cold that we are seldom without fire." His most urgent priority was "3 pair of good stout boots with thick soles" but he also asked him to send "my abbreviated dictionary of the Spanish Academy." There were plenty of shops in the town, but the owners all complained that business was bad. Food supplies left much to be desired: "Poultry is not to be had, nor eggs, vegetables are scarcely to be seen, pumpkins and corn which are brought in by the Indians completing the list. Milk is worth two reales the quart." Wage levels were also high:

> Servants' wages are very high owing to the price paid by the Company; my two men are now earning twenty dollars a month and their meals. Miners are paid a dollar a day, labourers, mostly boys of 14 or 16 four reales, masons three dollars a day, carpenters from 1½ to 2 dollars a day.[43]

As the population continued to increase, the newcomers had to be accommodated and we have two good contemporary descriptions of the houses. The first is from a letter sent to a newspaper by "a gentleman of the United States, residing in Chihuahua" who visited Guadalupe y Calvo in 1837:

> I went over at Christmas to see them. They are right in the heart of the Sierra Madre, and probably twelve to fifteen thousand feet above the level of the

> ocean. So we had a pretty cold time of it. But the British officers have built comfortable log-houses, and have plenty of wood to make fires. We spent a pleasant time, and remembered our friends at home. It is an English company but the greater part of the officers are Americans, and well paid.[44]

A few years later, the American traveler Albert Gilliam also visited Guadalupe y Calvo and he noted:

> The houses of Guadaloupe [sic.] are covered with shingles, and the windows had sashes and glass in them, as the English were the first builders there, and besides, timber is plentiful in the mountains.[45]

It also seems that the British miners were quickly integrated into the local community. For example, Edward Clark married Guadalupe Ordoñez and they had a son in 1846. Stephen Buchan, Robert Auld, and Charles Birkbeck all acted as godparents to children of Mexican couples.[46] Agustin Meinecke, a German miner, lived with "a young girl of rather decent family." Robert Auld's wife strongly objected to this arrangement, making "a great noise about it," and Meinecke was threatened with dismissal if he did not return the girl to her family. As the Mexican parents had no objections, the matter was eventually resolved, their daughter continuing to live with Meinecke.[47]

Auld's wife may also have disapproved of the lifestyle of John Hitchcock Buchan. He seems to have moved in with or set up house with Florencia Yañez, not long after his arrival. They had their first child, Joseph Robert Buchan, on 3 June 1839. Three more children followed in quick succession—Justo Stephen born 1 August 1840; John George, born 20 September 1841; Maria Clara Mauricia, born 22 September 1842. A fifth child—Mariana—was born to Florencia on 19 April 1844 but the parish register entry does not name the father although it may well have also been Buchan. There is little further information about Buchan's Mexican-born children or the future life of their mother, Florencia Yañez. Maria Clara Mauricia died in 1843 at just a year old and given the absence of any later reference to him, it is probable that John George also died in infancy. Two of the sons—Robert and Stephen—survived and after a brief period in Britain emigrated to Australia in 1858. Robert died in a drowning accident in 1860 but Stephen went on to establish a family that continues to this day. Buchan's own future career is explained in later pages of this work.

It is clear that the town and the mining industry with all its ancillary activities made rapid progress after the arrival of the British miners and the formation of the Guadalupe y Calvo Mining Company. The progress and rising silver production did not go unnoticed elsewhere. We do not know how the Mexican and South American Company of London became aware of the potential investment opportunity, but what is certain is that in May 1837 it began to invest in Guadalupe y Calvo and in neighboring regions. Again

Auld's younger sister, Katherine Carmichael Auld. Just four weeks later on 30 October, Robert Auld died at Guanaceví, aged forty.[52] Aware of the loss of his friend and now deceased brother-in-law, Buchan and his new wife, who was already pregnant in December 1846, decided to return to Mexico, where they arrived early in January 1847. By March, they had made the long and dangerous journey several hundred miles overland on horseback to Durango and from there to Guanaceví. From letters that Buchan wrote to Ewen Mackintosh in Mexico City, it is clear that he had been given authority by the Mexican and South American Company to continue working the mines at Guanaceví. Writing from there on 23 April 1847, he told Mackintosh that he had been authorized to spend a further £15,000 or 80,000 pesos and that Bendall and Co. of Zacatecas had already advanced him 30,000 pesos. A week later, he wrote again on notepaper headed, *Negociación de Guanasevi.*[53] Two other companies, Kerrison and Co. and J. B. Penny and Co., had now advanced him the remaining 50,000 pesos. Mackintosh who was acting as the Mexican and South American Company's main agent in Mexico, arranged the credit transfers between London, Mexico City, and Zacatecas.[54]

The last letter Buchan wrote to Mackintosh was from Durango on 30 August 1847. He referred to a letter he had received from London that told him of another draft for £6,000 that had been sent.[55] It appears, therefore, that the Mexican and South American Company intended to continue its investments in Guanaceví, but a short time later the directors decided to liquidate all their interests in Mexico. We do not know their reasons but there are various possibilities. Buchan had decided not to continue working the mines at Guadalupe y Calvo. The original contracts with the local owners had ended and they had demanded what one resident described as "exorbitant terms" for their renewal.[56] War between Mexico and the United States had started in 1846 and by 1847 the American forces were fighting their way to Mexico City. Although Buchan wrote that he did not expect to see the Americans at Guanaceví because it was not on their route, the war did have the direct effect of reducing the labor supply in the mining communities as volunteers and conscripts joined the Mexican army. The price of labor increased significantly as a result of the shortage. Finally, there was the constant threat of attacks by the *bárbaros*. In his letter to Mackintosh on 30 August, Buchan told him of a recent and unexpected attack early one morning by about 200 Indians. They had pillaged houses, stolen cattle, and carried off children, "amongst them a son of one of the first families." Perhaps surprisingly, Buchan made no mention in this letter of 30 August 1847 that on that same day, his wife Katherine gave birth to a daughter who was baptized Maria Isabel in Durango cathedral on the following 7 October.

The directors of the Mexican and South American Company, therefore, decided late in 1847 to leave Mexico and concentrate their attention and money on the copper mining industry of Chile. Their Mexican investments had not produced very large amounts but enough to pay the shareholders regular returns. The capital of the company was £200,000 in 20,000 shares of £10 each on which £7 per share was paid. The first dividend paid on 1 January 1837 was 10s. per share. A further eight dividends of 10s. a share were paid each year—two in 1838—until January 1844. A total of £45,000 was paid out in dividends between 1837 and 1844. Thereafter, dividends of various amounts were paid until the company was liquidated in 1859. There are no extant shareholder registers and hence for the most part, we do not know who the shareholders were. It is probable that most of the shares were held by the directors themselves but Finlay, Hodgson, and Company, a merchant firm in London, also had substantial holdings, either on their own account or on behalf of clients. They bought sixty shares in 1837 and also acquired 1,395 shares for Messrs. Muriel, "their correspondents in Paris."[57]

As for the British miners, blacksmiths, carpenters, doctors, clerks, managers, and others who had hastened to Guadalupe y Calvo when news of the discovery of the silver and gold deposits had spread in 1836, most seem to have left to seek their fortunes elsewhere. Buchan returned to Real del Monte where in May 1848, he was appointed the senior commissioner of the British Real del Monte Company. The following year, when the British company was sold, he joined the Mexican-owned business that was formed to work the mines and became one of its shareholders. He stayed with the new company until 1856, when he retired and returned to live in Britain together with his wife Katherine and their family of six children, two born in Guadalupe y Calvo, one in Durango, and three at Real del Monte. Others went to Guanajuato and Zacatecas to work for the British companies still operating in those areas and some, for example, Edward Coffin, joined the gold rush to California. A few Britons, however, chose to remain in Guadalupe y Calvo, for example, Thomas Lambie, a carpenter; Frederick Evans, a miner; Charles Bowring, geographer and miner; and Joseph Scobie, Richard Killaly, and Matthew Wigham, who all worked for the local Mint. Finally, there was William Kays who made his living as a tailor at Guanaceví.

EQUITABLE MEXICAN MINING ASSOCIATION

In 1835, a prospectus was issued announcing the formation of this new company. It was to have a nominal capital of £50,000 in 2,000 shares of £25 each, with the first deposit being £2 10s. The intention was to work "an undoubted rich and important gold mine." No copy of the original prospectus has been found and *The Quarterly Mining Review* that reported its appearance noted that very few details were given to prospective shareholders. Names of the directors or promoters were not given and the only person identified was the company's solicitor, Mr. George, Warnford Court, Throgmorton Street. Also, very unusually, there were no advertisements in the London or the provincial press. No other reference to the company appeared and it is likely that lack of interest or response to the prospectus caused it to be abandoned.

THE PACHUCA MINING COMPANY

In 1844, another British-backed Mexican mining venture was announced. This was the Pachuca Mining Company. The town of Pachuca is located just a few miles from Real del Monte, where the British Real del Monte Company continued to be active in its working of the district's mines. Always anxious to try out new mining techniques, especially in the difficult area of refining the ore, the company had been experimenting with what was known as the barrel or Freiburg method of amalgamation. The results of the experiments appeared to demonstrate that certain types of ore known as rebellious or refractory, from which it had proved impossible to extract silver profitably, could be made profitable with the new refining process. It was known that this low-grade ore was abundant in mines around Pachuca. Some of the directors of the Real del Monte Company were persuaded that it was worthwhile acquiring the rights to these Pachuca mines and they decided to set up a company to work them. The new Pachuca Mining Company was formed around May 1844. Its capital was £20,000 in 2,000 shares of £10 each with a deposit or first installment of £2 a share. The company was immediately registered and its shares quoted on the Stock Exchange. The mines of La Esperanza and La Regona had been contracted at Pachuca.[58]

In May 1845, the first annual general meeting was held in London under the chairmanship of Sir Robert Price and with John Phillips as the company secretary. Shareholders were given an optimistic picture of the work in progress as well as the costs so far incurred. 13,000 pesos had been spent on clearing the old mines and more productive ground was now being worked. Expenses in acquiring the rights, directors' fees, legal and other costs had taken £2,410.

For the next few years, reports of the Company's progress were published regularly in *The Mining Journal*. The mine manager at Pachuca was Captain Trenear. At least three members of the Cornish Trenear family—John, James, and Henry—are listed in the passport records, all residing at Real del Monte, and it is not clear which one was involved with the new Pachuca Company. William Rule, also a long-standing employee of the Real del Monte Company, was the director at Pachuca and he also sent reports back to London. At the third annual general meeting in May 1847, Sir Robert Price again reported on expenditure rather than income. In 1846 12,019 pesos had been spent on further trials but the most productive mine, La Rejona, had only yielded a few hundred pesos after the costs of production and refining had been met. Trials were continuing and there were some promising signs.

It seems that poor results and shareholder disappointment brought the Pachuca Mining Company to an end a few months later in 1848. £4 had been paid on each £10 share meaning that £8,000 or 40,000 pesos had been raised from investors. While insignificant in mining terms, that expenditure of £8,000 had brought no returns. The market price of the shares did not rise above £4½ and profits were never enough to pay a dividend. Another factor in the winding up of the Company was the sale of the British Real del Monte Company to Mexican investors at about the same time.[59]

THE CERRO DEL BOTE MINING COMPANY

The El Bote Company, as it became known, had an unusual and in many respects intriguing history. The main characters in the story were two young men who, like so many others of their age, sailed for Mexico in the 1820s to make their fortunes. We know little of the first of these, Gennaro Placci. He was of Italian origin, born in 1811, probably in Florence, but with a British passport. He arrived in Mexico in 1826 and obtained employment with the Bolaños Company at Zacatecas. The following year, 1827, Roger Allday Kerrison arrived to work for the Real del Monte Company at Pachuca and Real del Monte. He was born in 1803 at Norwich, the younger son of a very prominent banking and landowning family. Given his family's personal connections throughout the Norfolk region, it is certain that he would have known the mining engineer John Taylor, who had been the principal promoter of the Real del Monte and Bolaños Companies. The Kerrisons were also related to the Harveys through marriage and business, and Roger would have known Sir Robert Harvey and most of the other major shareholders in the two companies. After a few years at Pachuca, he moved to Zacatecas to work for the Bolaños Company and it was there that he met Gennaro.

Both men continued to work in Zacatecas for several years. They managed mines for various British companies and by 1836, Gennaro was employed as the chief administrator of the Bernardez refinery, a property belonging to the Mexican-owned Fresnillo Company.[60] Then, in June 1837, he arrived in London and a few months later, in April 1838, he published a pamphlet.[61] He explained that during his time in Zacatecas, he had become interested in the mining district known as El Bote. This was located about two miles from the city of Zacatecas. It was known to have substantial deposits of gold and silver but local Mexican miners, lacking sufficient capital, had only engaged in small-scale surface mining, with poor results. The large and well-capitalized British mining companies had shown little interest, preferring to concentrate their efforts on restoring the better-known and more established mines. Seeing, therefore, a very promising opportunity, Gennaro had persuaded local miners to sell him their mining rights in the area. He proposed to set up a new company to be called the Cerro del Bote Mining Company. It was to have a capital of £100,000 raised by the sale of 1,000 shares of £100 each, with the first payment being £10. Seeking support for his claims of their potential, he had sent "the whole of the plans and documents relative to the Cerro del Bote mines" to John Taylor. His reply is included in the pamphlet. He strongly supported the proposal concluding that "I am satisfied that the choice of the ground is a good one, and that in working it there would be every reasonable prospect of success" (p.18). Furthermore, Gennaro added, several prominent mining engineers working in Mexico, notably Joseph Burkart and Ferdinand Schuchardt, had already requested shares in the new company. Letters from them together with other documents explaining and lauding the immense potential of the mines are included in an appendix to the pamphlet.

The first appearance of the Cerro del Bote Mining Company, therefore, was in London in April 1838.[62] There is no further mention of it after that time and it seems that Gennaro's persuasive prospectus failed to attract investors. Doubtless disappointed by the lack of interest in his scheme, he returned to Zacatecas, where for the next few years he continued his career in silver mining. By 1844, he and his friend Roger Allday Kerrison were "well-established citizens of the Zacatecan community," both members of the important regional Mining Tribunal, the body charged with regulating the mining industry in the state.[63] Gennaro is described in the passport records as a merchant and by the early 1840s, he may have been engaged in both mercantile and mining business. Roger Allday also seems to have added commerce to his mining interests. Probably in association with George Kerrison, a cousin or near relative, he set up the partnership of Kerrison and Company of Zacatecas. It was engaged in various mercantile and financial transactions in the early 1840s, corresponding regularly with Manning and Mackintosh in Mexico City.[64] Finally, in 1845 Gennaro and Roger Allday decided to form a partnership and to try once again to set up a company to finance the working of the Bote mines. They agreed that Roger Allday should leave for Britain at once in search of investors.[65]

As we have seen earlier, the Bolaños Company leased the Bote mine in late 1845 or early 1846 from its proprietors, Kerrison and Placci. Details of the agreement are lacking except that it included an annual *alimento* or payment of £3,000. With the agreement in his pocket, Roger Allday returned to Mexico. Hopes of a bonanza from the Bote mine, however, were soon disappointed, and unable to persuade the shareholders to provide more funds, the directors of the Bolaños Company decided to dissolve the company. This decision was taken around late November or early December 1849 but it was not as clear or uncomplicated as it first appeared.[66] On 5 December a special meeting of shareholders was summoned "by circular" to meet in the London Tavern. The purpose of the meeting was to discuss the formation of a new company to be called The Cerro del Bote Mining Company with a capital of £50,000, in 5,000 shares of £10 each. John Taylor Jr., son of the long-time manager and principal promoter of the both the Real del Monte and Bolaños Companies, explained the situation. The Bolaños Company, having failed to raise the capital required to work the Cerro del Bote mines, had reluctantly returned them to their owners. They had agreed, however, "to enter into a fresh contract" for the mines. In return for a payment of £20,000 paid to them in London, they would transfer all the mines and the reduction hacienda of *Cinco Señores* to the new company that it was now being proposed to establish. The stock and property included "two steam engines (one erected and in full work on the mine), stores of quicksilver, forage, fuel, timber, iron, steel, tools, cattle, implements, and ores on hand; the whole of which are indispensable, and could not be placed thereon for three times the sum named above." Future profits would be divisible between the company and the owners, who were to receive in advance an *alimento* of £2,400 per annum. Shares to the amount of £15,000, it was announced, had already been taken.

The Cerro del Bote Mining Company had been provisionally registered on 1 December 1849. The registration document reveals that the promoters were John Taylor Jr. and Roger Allday Kerrison. Roger Allday's stay in Mexico where he had arrived in January 1847 had been very brief and knowing that the Bolaños Company was running out of money, he had once again crossed the Atlantic to find investors for his Bote mine. Advertisements announcing his new venture were placed in the newspapers in December 1849 and by the end of that month separate notices were published advising investors to make early application as the share subscription list was soon to be closed. In fact, things were not going as well as Taylor and Roger Allday had hoped. *The Mining Journal* reported on 12 December 1850 that the subscription lists were progressing very slowly with only £22,000 of the £50,000 capital so far raised. A few days later, John Phillips, for many years secretary of the Real del Monte and Bolaños Companies, tried to give the shares a boost. In a letter published in *The Mining Journal* on 19 January, he argued that the new company had a very promising future. Many of the former Bolaños Company shareholders, he wrote, including the chairman, Sir Robert Price, had taken shares.

On 16 February 1850, *The Mining Journal* reported that as only £25,000 had been subscribed, the attempts to form the Cerro del Bote Company were being abandoned. Once again, the hopes of raising new capital for investment in the Mexican mining industry were unsuccessful. It was not surprising.

California now offered better prospects for investors with its rapidly expanding gold mining industry, and many new companies were being formed to develop mines in Australia. Also, by this time, with just one exception, all the main British mining companies had either been dissolved, or were in difficulties and facing the imminent prospect of being wound up. The majority of the Bolaños Company shareholders evidently were not prepared to risk even more of their money, notwithstanding the prospects and promises of future dividends held out to them by John Taylor, Jr., John Phillips, and the editor of *The Mining Journal*, who expressed his surprise that the Cerro del Bote Company had failed as he was sure it would have yielded "a remunerative profit."

Roger Allday Kerrison may have been disappointed. As owners of the Bote mines, he and Gennaro Placci would have received the £20,000 from the new company, as well as the yearly *alimento* of £2,400. In fact, although he was not to know it at the time, the failure to set up the new company as a joint stock enterprise turned out to be extremely beneficial to him and to his descendants for the next hundred years. He decided to return to Mexico to work the mines under the control of his own privately owned Cerro del Bote Mining Company in which he and Gennaro would be the main shareholders. He left London almost immediately, arriving back at Zacatecas in April 1850. There are few details available of how the mine was developed in the 1850s but it appears to have been profitable almost from the beginning. Cross provides some useful statistics of the value of the silver produced:

1856 274,400 pesos
1857 219,275
1860 204,485
1862 155,418
1871 296,106
1873 300,920[67]

While these amounts were less than some of the other Zacatecas mines operating at the same time, they were nevertheless impressive, confirming that the mine had turned out to be an excellent investment. A British traveler, Charles Sevin, who visited Zacatecas in 1856, wrote that "the Bote of Zacatecas has given splendid profits to two or three of the English owners."[68] One of the managers was Rayner Hector Henry Alexander. Born in Northumberland, he had worked for other British mining companies and in later years was a merchant in Zacatecas as well as British vice-consul. In 1851, he took the trouble to send an ore sample from the Bote mine to the Ipswich museum that, unfortunately, has since disappeared or cannot be located. Later managers included Archibald MacGowan and Henry Trenear. Among the workforce were "sundry Cornishmen," possibly recruited directly from Cornwall or more probably from the British companies like the Bolaños Company that were winding down their activities. The 1850s were a time of great upheaval in Mexico's domestic politics with a major civil war raging for several years. Forced loans, additional taxes, illegal confiscation of silver, and seizure of livestock and supplies by marauding bands of rebel soldiers were commonplace and the El Bote Company suffered considerable losses in this way. In a British claims commission of 1862, the Company lodged a claim for 63,431 pesos.

Roger Allday Kerrison stayed in Zacatecas for a few more years until around 1860 when he returned to Britain and Norfolk where he settled down as a retired country gentleman. He and his brother, also Roger, were partners in the Hudson and Harvey Bank, otherwise known as the Crown Bank, of Norwich. They were apparently very largely silent or sleeping partners in the bank, taking no part in the daily management of its business. It came, therefore, as a major shock to them, and to the whole of the Norfolk élite, when in July 1870, the bank was declared bankrupt. The director or manager of the bank was Sir Robert Harvey, son of the founding member of the same name of the Bolaños and Real del Monte Companies in the 1820s. Harvey, it transpired in the subsequent enquiries, had long been gambling with the bank's funds and in the early months of 1870, he had lost as much as £600,000. He committed suicide in July 1870. His two partners, the Kerrison brothers, although they had taken little or no part in the affairs of the bank, were liable for its debts. All their personal property was taken by the bankruptcy court and sold in public auction.[69]

The details of the bankruptcy and sale of the Kerrison assets need not concern us except in one respect. Included in the sale of Roger Allday's property were his shares in the El Bote Company and in other Mexican mines. The notice advertising the sale appeared in *The Daily News*, 6 February 1873. It described his Mexican investments as follows:

> the shares in the valuable and extensive GOLD and SILVER MINES, Mining Plant, Machinery, and Materials in the City of Zacatecas, Mexico, with the Stores and Utensils as now used and carried on in the Mining Business of the Bote Negotiation. The share of Mr. Kerrison in the above mines and property consists of:

Six Twenty-fourth Parts or *Barras* in the Bote Mine.

One Fourth of 18-24th Parts or *Barras* in Los Cleregos Mine.

Six Twenty-fourth Parts or *Barras* in the Hacienda de Cinco Señores (Reduction Works).

It appeared, therefore, that the Kerrison interest in the Bote mine had come to an end. The shares were sold in the auction held on 25 February at the Mart, Tokenhouse Yard, London. At a creditors' meeting held a few months later in Norwich, the bankruptcy trustee's accounts were presented, and they included £500 from the sale of the "Mexican mines." One of the creditors present asked why the mines had only raised £500 to which the trustee replied that "that was all they could get for Mr. Kerrison's share."[70] In fact, what appears to have happened was that William Henderson, Roger Allday's brother-in-law, had offered £1,000 for the mining company shares but the bankruptcy trustees refused the offer, believing that they were worth much more. They sent a man to Mexico to evaluate the mines and other properties but on his return, he reported that they were of low value, having paid little or no dividends in recent years. The trustees decided to accept the offer that had already been made but Henderson now reduced it to £500. That was accepted and Henderson generously returned the shares to Roger Allday. Within a year, the mines had returned to profitability and again produced handsome dividends, as Roger Allday's son recalled in his autobiography: "In the first year after the bankruptcy, the mines recovered and my father made a profit of £4,000 out of which he repaid my uncle the £500 he had lent him."[71]

On 2 August 1882, a few months before his death on 31 October 1882, Roger Allday signed his will. It makes interesting reading. Among his many bequests from an estate of £34,094 were eight shares or *barras* in the Bote mine. Mentioned more than once in complex provisions, he bequeathed four shares to a family trust that he asked his executors to establish. The main beneficiary of the trust and therefore of the Bote shares, after a period of some years, was to be his younger son, Edmund Roger Allday. He also stipulated that the management of the El Bote Company should be entrusted to his elder son, Roger, to whom he also left four shares. Among personal items to be treated as family heirlooms were "the large silver centrepiece with a mule and a man blasting rocks, and the large silver cup which was presented to me in the year one thousand eight hundred and thirty by the officers of the Veta Grande Negotiation, and my silver coffee and tea service, and my six napkin rings, and my three silver mugs made of silver from the Bote mine." Finally, he left £250 to Adelaide Kerrison Placci, "daughter of my late friend Gennaro Placci."

Shares in the Bote mine, therefore, remained in the Kerrison family for another generation. Few details of the company's management in the latter quarter of the nineteenth century and first decades of the twentieth are known and no company accounts, financial or otherwise, have been located.[72] In his autobiography, Edmund Roger Allday Kerrison refers several times to his attendance at "Bote meetings" but he gives no details. Mr. and Mrs. Roger Kerrison, presumably referring to Roger Allday's son, visited Mexico in 1875 and it may be assumed that they went to Zacatecas to check on their Bote mine. Other members of the Kerrison family, for example Edmund Roger Allday, also made the long and hazardous journey to Zacatecas to check on their property. In 1898, the death of James G. Wood was announced. He had been, it was said, a very successful manager of the Bote mine for many years. The mine, the report added, belonged to "a syndicate chiefly residing in Ipswich." [73] He was replaced as general manager in 1900 by J. S. Pattinson, from Liverpool, and by that time, there seems to have been a sizable British community in the workforce. Part of the reason for the paucity of information about the company may be explained by a comment made by an American visitor to Zacatecas in 1887. In an article headed "The Britishers' Good Thing," he wrote that the Bote gold mine was controlled by an English company that was very secretive about its operations, maintaining "a guard of reticence that is seldom penetrated." The bullion produced by the mine was shipped directly to England and one employee had revealed that in 1885 and 1886, 14,000,000 dollars in gold had been sent. A trip to the bottom of the mine, he added, was "a muscle-straining descent of ninety ladders."[74] Another American, seeking to promote a new mining company in 1891, emphasized that his mining property in Zacatecas had the same geological formation "as that of its famous neighbour, El Bote, which has been in bonanza over thirty years, and has given its lucky owners a million [dollars] yearly in dividends."[75] Such reports by visitors to Zacatecas may have exaggerated the success of the mine, but profits were certainly made. For example, dividends paid per share or *barra* in the 1870s were as follows:

1873: £150
1874: £250
1875: £200
1876: £200
1877–1878: £250

The Kerrison and Placci families, owners of most of the shares, were therefore receiving an annual income of several thousand pounds from the Bote mine.[76] In 1896, a Mr. W. E. Godfrey, described as one of the richest mine operators in Mexico, commented that the Bote mine "has been declaring dividends for fifty years, all of which have been paid to an English family in London."[77]

There is no doubt, therefore, that the Bote mine proved to be a very profitable venture for the Kerrison family, but the exact nature of their participation in later years is not entirely clear. There were several reports that the mine had been sold to American investors. The *New York Times* (28 September 1893), for example, reported that the gold mine of the El Bote vein had been sold to "capitalists from Richmond." In 1906, it was reported that the mine was being sold to an American syndicate and "the reason that it is to be sold is that those who now own it are less interested in mines than in other industries to which their wealth, gained from this mine, has absorbed their interest."[78] Press reports claimed that a sale price of $3,000,000 had been agreed.[79] No sale seems to have been concluded because three years later, there is evidence that the Kerrison family retained its interest and control of the mine. In London in 1909, there was a court case involving Kerrison v Glyn, Mills, Currie, and Co. Highly technical in many respects, the essence of the case concerned the sum of £500 which Roger Kerrison had asked his bank in Ipswich to transfer to a bank in New York. What is of interest to us are the details that were presented to the court concerning the Bote mine. In 1903, in what the judge described as in effect a revolving credit, Roger Kerrison, the "managing owner" of the El Bote Mining Company, established a procedure with his bankers whereby £500 was periodically sent to a bank in New York for the use of the mining Company.[80] Its managers were thus able to write cheques on the account in the New York bank. Some time afterward, Roger Kerrison transferred the management of the mine to his son, Roger Orme Kerrison. He continued the revolving credit but in October 1907, just as £500 was dispatched, the New York bank was declared bankrupt. Roger Orme requested the return of the £500 and the British bank involved—Glyn, Mills, Currie, and Co.—refused his request.

The details and outcome of the case need not concern us.[81] What is to be noted is that the Kerrison family still owned the Bote mine but now it appears that they were subsidizing its operations, presumably because it was not profitable. Even so, very substantial investments in new equipment and machinery were still being made. In 1908, the El Bote Company was one of the first to install a new cyanide processing plant with the capacity to handle 300 tons of ore per day. At the same time, large machinery orders for powerful gas engines were placed with a German manufacturer.[82] Again details of the immediate effects of such innovations on the business are lacking, but the Kerrisons were still involved. Quarterly dividends of £200 or more producing up to £1,600 a year were regularly received throughout the 1920s, and at least some of this money was invested in various British government stocks. By 1921, the general manager was Roger Fulke Kerrison, son of Edmund Allday and grandson of the founder, Roger Allday. He remained for several years in the 1920s and after he gave up the post still made periodic visits to Mexico. The mine is also mentioned in Roger Kerrison's will as amended in 1924. His son, Roger Orme, accompanied by his wife, had visited Zacatecas in 1909. He died on active service in the First World War. The family shares in the Bote mine, therefore, were left to John Orme Kerrison, the testator Roger Kerrison's grandson.

There is one more aspect of the El Bote Mining Company history to mention, and that is the Placci family connection. Gennaro Placci continued to work in Zacatecas until 1856 when he decided to escort his growing family back to Europe. With his Mexican-born wife, María Guadalupe Ruíz Villegas, and five young children, he sailed from Veracruz in April of that year, traveling to Britain via New Orleans and New York.[83] He spent some time in Britain, visiting his friend Roger Allday, whose bank account shows that he made several payments to him in June and July of sums varying from

£20 to £45.[84] Their final destination was Italy and although Gennaro returned to Mexico for one more brief visit, he soon returned to Europe, first to Britain and then to Florence, where he died on 23 May 1863.[85] His children and grandchildren retained their close friendship with the Kerrisons, and for the rest of the nineteenth century, they were present at Kerrison family weddings and other events. In 1893, for example, Adelaide Rosa Kerrison was married and among the guests were Madame Placci, whose wedding gift was a diamond and sapphire bangle, and Mr. G. Placci, who gave the newlyweds silver gilt sugar tongs and spoon.[86] By that time, Mr. Gennaro Placci had been appointed the British vice-consul in Florence. It is also worth noting that among the wedding gifts was a silver salver presented by the "Officers of the Bote Mine." Another connection between the two families was evident in January 1909. A major earthquake struck Italy in 1907 and the Lord Mayor of London set up a fund to help the victims. Among the donors with a donation of £50 were "Messrs. Kerrison and Placci."[87] Finally, in the Mexico City English-language newspaper, *The Mexican Herald* (23 September 1906), there is a report that confirms the Kerrison/Placci connection. "The Bote Mine," it says, "is owned by a company composed of four men, the descendants of an Englishman and an Italian who discovered the veins seventy years ago, and have developed the property to the present point. The mine has paid big dividends for sixty or more years, and has had many famous bonanzas." The "four men" referred to were the two Kerrison brothers, Roger and Edmund Allday, and two of Gennaro Placci's family. His widow, María Guadalupe, had signed new articles of association for the company on 2 January 1888 and, like the Kerrisons, her children inherited the family's shares in the Bote mine.

The El Bote Mining Company was finally wound up in 1938, one hundred years after Gennaro Placci had first arrived in London in search of investors. We do not know their reasons, but on 9 July 1938, the shareholders resolved to close the company, liquidate all its assets, and divide the proceeds between them. The Bote mine had been sold, almost certainly to what was named the Carnegie Bote Mining Company. The proceeds were divided among the shareholders, and London accountants were instructed to sell off all the company's other assets. These consisted of a large portfolio of shares in more than forty companies including Rolls Royce, Shell Oil, and the National Railways of Mexico. Together with cash in hand in the Bank of Montreal and Barclays, the accounts show that the total amount raised was £67,185.2.2. Interestingly, they also reveal that the Kerrison and Placci families owned nineteen shares or *barras* in the El Bote Company. The asset sale total of £67,185 provided a payment per share of £3,425. The shareholders who received this very substantial payout were as follows:

John Orme Kerrison (8 shares) £27,923. 18.4.

Edmund Roger Allday Kerrison (2 shares) £6,980 19.7.
Roger Fulke Kerrison (2 shares) £6,980 19.7.
Carlo Placci (3½ shares) £12,627 18.1.
Gennaro Placci (2½ shares) £9,019 18.8.
M. Henraux (1 share) £3,607 19.5.

The Kerrison family, therefore, had the majority holding with twelve shares and the Placcis held seven. Although nothing is known about his involvement in the El Bote Company, the M. Henraux listed in the accounts was probably Lucien Henraux. He was a relative of the Placcis, who may have given him his one share.[88]

As we have noted earlier, ownership of Mexican mines was always divided into twenty-four shares or *barras*. The Kerrisons and Placcis/Henraux had nineteen, but with one exception, the identity of the owners of the remaining five is not known. The exception was Samuel Bradford Birkbeck. Accompanied by his younger brother, Charles, he arrived in Zacatecas in January 1827. He soon found work and over the next twenty years, he managed mines and refining plants for the Bolaños and other British mining companies. He also developed his own business interests independently and had many transactions with Roger Allday Kerrison, Kerrison and Company, and Gennaro Placci. He invested in several mines in the Zacatecas area and by the late 1840s, his journal reveals that he had some financial interests in the Bote mine, contributing regularly to the *alimento* payable to the mine owners.[89] His exact participation at this point is not known but when Roger Allday Kerrison and Gennaro Placci set up the El Bote Mining Company in the spring of 1850, Birkbeck obtained at least one *barra* in the new venture. He mentions it several times in his journal and in the entry for June 1854, he notes that he had received his first dividend of 500 pesos.[90] He continued to collect dividends in following years and Plomley has calculated that between 1849 and 1862, he received £4,855 17s.11d. His investment was so profitable "that Birkbeck decided to retain a share in the mine even after his departure from Mexico in 1860."[91] His brother, Charles, was apparently manager of the Bote mine in the 1850s and he may also have been a shareholder. Finally, Plomley suggests that the British merchant William Newall, who had extensive merchant interests in Zacatecas as well as mining investments, may also have been involved in the El Bote Mining Company.[92]

Birkbeck left Mexico for the last time in 1860, taking his Mexican-born wife, Damiana de Barre Valdez, and their eight sons and one daughter to settle in Australia.[93] He bought land near Rockhampton, Queensland, and together with his Mexican servant, Denis Cifuentes, he developed a large rural estate. He died in 1867 and his share in the El Bote Mining Company and in other Mexican mines passed to his heirs. Disputes arose over the payment of dividends in the 1870s but their agent in Zacatecas, Frederick

Palmer, urged them to keep the Bote share. It was worth between 20,000 and 40,000 pesos, he told them. Furthermore, "the mines are in a splendid condition" and "each *conducta* regularly takes from 40,000 to 60,000 dollars."[94] Further research in the extensive Birkbeck archive may reveal the eventual fate of his single *barra*. It is possible that it was sold to Roger Kerrison in 1881. Plomley notes, without explanation, that "one exceptionally large amount, more than £11,000," was received from Mexico in that year.[95]

The complete history of the Bote mine and the British mining company that worked it has yet to be written, but the short summary given here already reveals some exceptional features. Mexico's development from 1850 to the 1930s is characterized by foreign aggression and invasion, civil war, dictatorship, and revolution. Diplomatic relations with Britain were suspended for almost twenty years between 1867 and 1884, following British involvement in the events that brought about the French invasion and creation of the empire headed by the Austrian Hapsburg Prince, Maximilian. The Maximilian empire survived for three years between 1864 and 1867, and after its defeat, Mexico soon reverted to internal conflict and dissension. Only the rise of the dictator Porfirio Díaz brought about political stability and economic progress. The *porfiriato* or Díaz dictatorship lasted from 1876 until 1910, when the so-called 1910 Revolution began. One of the major events in world history, the 1910 Revolution caused immense destruction and loss of life in many parts of the country. Even though, as Professor Knight has noted, in general, foreign-owned mining interests were not "conspicuously objects of nationalist or xenophobic attacks," there were instances of forced loans and confiscations at the local level.[96] In just a single example of a forced loan levied in Zacatecas by a military commander, in 1913 the El Bote Mining Company was forced to pay 5,000 pesos.

Against this historical background, the survival of a foreign-owned mining company is in itself noteworthy, but adding in the fact that the company was privately owned by a British family of Norfolk and an Italian family residing in Florence, the achievement is remarkable. Throughout all the upheaval and change, the Kerrison and Placci families had managed to retain their possession of the Bote mine, deriving handsome dividends from it, for exactly one hundred years.

The El Bote Mining Company was the last British mining company to be established in Mexico in our period up to 1859. It was, of course, certainly not the end of British interest in Mexican silver and gold. Political events, civil war, and foreign invasion seem to have been no deterrent to British investors. In the decade between 1860 and 1870, at least ten more companies were announced in London that intended to exploit newly discovered or old mineral deposits. Glowing prospects of high returns were offered to those who bought shares in the Pachuca Mining Company (capital £10,480 in 10,480 shares of £1 each) with mines at Pachuca; the Santa Elena Silver

Mining Company (capital £30,000 in 15,000 shares of £2 each) with mines in the Real del Monte district, fifty-five miles north of Mexico City; the El Chico Silver Mining and Reduction Company (capital £75,000 in 15,000 shares of £5 each) with mines at El Chico, nine miles from Pachuca and seventy miles north of Mexico City; the Laguna Silver Mining Company (capital £20,000 in 4,000 shares of £5 each) also with mines at El Chico; the Capula Silver Mining Company, also at El Chico (capital £50,000 in 25,000 shares of £2 each); the Almada Silver Mining Company (capital £75,000 in 7,500 shares of £10 each) with mines in the northern state of Sonora; the Guerrero Gold Mining Company (capital £50,000 in 50,000 shares of £1 each) with mines and "auriferous alluvial deposits" in the state of Guerrero, about 240 miles southwest of Mexico City; the San Pedro del Monte Silver Mining Company (capital £75,000 in 7,500 shares at £10 each), with mines at Ixtapan del Oro, about two days' journey by road from Mexico City; the Sonora Silver Mining Company (capital £100,000 in 40,000 shares of £2 10s. each) with mines in Chihuahua and Sonora; and finally, the New Rosario Silver Mining Company (capital £50,000 in 50,000 shares of £1) with mines in the Real del Monte district. These companies provided a total new investment of £535,480 with 231,980 shares offered to the British public. The same decade also saw the foundation in 1864 of the British-owned Bank of London and Mexico and the formation in London of the Imperial Mexican Railway Company with a share capital of £2,700,000 in 135,000 shares of £20 each.

We have now identified fourteen British companies formed between 1821 and 1859 to exploit Mexico's rich deposits of gold and silver. The mining industry was clearly the main attraction for British investors, but numerous other enterprises, some financed at least in part from Britain, were also created by British entrepreneurs ranging from plumbing services, carriage manufacture, textile factories, and mints, to international merchant houses in several regions of Mexico. All manner of other services were offered by Englishmen from *haute couture* to language classes and there were few cities or large towns in the country where a Briton could not be found. Notwithstanding the company failures from the very early years of the British Mexicomania, the British investors never lost their faith in the future prosperity of Mexico. Events appear to have had little effect on their private assessment of risk, and there were always those willing to take a gamble with their money on shares in Mexican ventures.

NOTES

1. Letter signed by fourteen British companies to R. Pakenham, June 24, 1836, NA, F.O. 50/99, fols. 188–92.

2. Weichsel refers to his negotiations with the local *cura* in a letter to the directors dated Peñoles, August 1, 1831 published in the *Quarterly Mining Review* 2, no. 6 (January 1832): 366–67.

3. *Quarterly Mining Review* 1, no. 4 (March 1830): 545. A *carga* or load of ore was 300 Mexican pounds weight which equaled 304 English pounds, approximately.

4. The first half-yearly meeting was held on November 7, 1831: see notice in the *Times*, October 29, 1831.

5. Directors' report to meeting of shareholders on November 7, 1831 in the *Quarterly Mining Review* 2, no. 6 (January 1832): 370.

6. Archivo Histórico de Notarías de Oaxaca, escribano Vicente de Castilleros. Libro 160, fol. 21v.

7. Ibid., fol. 44v.

8. Muhlenpfordt's work on the Zapotecs was published as *Los palacios de los zapotecas en Mitla* (Mexico: UNAM, 1984). His more general work was published as *Ensayo de una fiel descripción de la República de México* (Mexico: Banco de Mexico, 1993).

9. The company reports were published separately but very few examples have survived. The National Library of Australia has one: *Report of the Directors of the Peñoles Gold Mining Association to Be Presented at the Fourth Half-Yearly General Meeting of the Proprietors Which Will Be Held at the Office of the Association on Monday, the 6th of May 1833* (London: E. B. and G. Clarke, 1833), 38pp. The British Library has the *Peñoles Gold Mining Association. Report Dated 4 May 1835* (London: T. Smith, 1835), 4pp. This includes brief details of the meeting held on November 3, 1834.

10. *Quarterly Mining Review* 2, no. 6 (January 1832): 363.

11. Quotations from *Report of the Directors of the Peñoles Gold Mining Company* (1833), 6, 7, 9, 11–12.

12. *Report of the Directors of the Peñoles Gold Mining Company* (1833), 15–16.

13. Archivo Histórico de Notarías de Oaxaca, escribano Vicente de Castilleros, Libro 162, fol. 27. Weichsel transferred this mine to the Company directors Messrs. Druke and Swinburne on April 5, 1833; Libro 162, fol. 105.

14. Archivo Histórico de Notarías de Oaxaca, escribano Vicente de Castilleros, Libro 162, fol.146.

15. *Times*, April 1, 1837.

16. There are several notarial documents concerning Sadler's business ventures in the Archivo Histórico de Notarías de Oaxaca.

17. In the 1842 census taken at Oaxaca there are other inhabitants said to be English but the names are unusual, for example, Andrew Fenochio and William Vafeet, *Padrón general de los habitantes de la Ciudad de Oaxaca* (Oaxaca: Instituto Nacional de Antropología e Historia, 1981).

18. The share certificate is number 302. I am grateful to Mr. Brian Mills for giving me a copy.

19. R. O. Auld and J. H. Buchan, *Notice of the Silver Mines of Fresnillo* (London: Cochrane and M'Crane, 1834), 20.

20. There are several receipts for dividends paid to Phelps in the Manning and Mackintosh Papers. The Manning and Mackintosh Papers are held in the Nettie Lee Benson Library at the University of Texas at Austin. I have used a microfilm of the collection made before it was fully catalogued and most of the documents have no reference number except a date.

21. AGN, *Gobernación. Siglo* XIX. *Relaciones Exteriores. caja* 33, *exp.* 395, fol. 142, passim.

22. AGN, *Cartas de Seguridad*, vol. 60, exp. 54, fols. 45–46.

23. Details from *Informes de la junta menor permanente de la Compañía de Minas Zacatecano-Mexicana* (Mexico: Impr. de I. Cumplido, 1837, 1838).

24. *Mining Journal*, April 14, 1838, 118.

25. *Mining Journal*, May 24, 1838, 162–63.

26. *Mining Journal*, February 22, 1840, 59.

27. The Bolaños Company held shares valued at 500 pesos in their accounts. See *Report*, June 25, 1844, Bolaños Company Reports (Bancroft), 2:293.

28. C. Richardson, *Mr. John Diston Powles: Or the Antecedents as a Promoter and Director of Foreign Mining Companies or an Administrative Reformer* (London: printed for the author, 1855), 12–13.

29. A John Henry Bodmer died at Merthyr Tydfil in 1856.

30. There are other versions of the discovery. I have followed the ones given in F. R. Almada, *Guadalupe y Calvo* (Chihuahua, Mexico: [s.n.], 1940), 33–74; Z. Márquez Terrazas, *Pueblos Mineros de Chihuahua* (Chihuahua, Mexico: Universidad Autónoma de Chihuahua, 2007), 135–39.

31. H. G. Ward, *Mexico in 1827*, 2 vols. (London: Henry Colburn, 1828: 2nd expanded ed., 2 vols., London: Henry Colburn, 1829), 2:580.

32. For details of this incident, see A. Probert, "Punishment of the Tizayuca Silver Wagon Bandits," *Journal of the West* 14, no. 2 (1975): 114–29.

33. *Morning Chronicle*, February 28, 1834.

34. Ewen Clark Mackintosh was born in London on May 16, 1802, the son of Ewen and Elizabeth Mackintosh. Henry Alexander was born on March 11, 1808. Both boys were christened at St. Martin in the Fields, Westminster.

35. B. A. Tenenbaum and J. N. McElveen, "From Speculative to Substantive Boom: The British in Mexico, 1821–1911," in *English-Speaking Communities in Latin America*, ed. O. Marshall (London: Palgrave MacMillan, 2000), 59.

36. Philip Baker managed the Veracruz branch until he resigned in 1840: see *London Gazette*, notices of partnerships dissolved in the *Manchester Times and Gazette*, September 5, 1840. Baker remained in Mexico for many more years, being appointed British vice-consul at Acapulco in 1850. Marshall formally resigned his partnership in 1842; see "Partnerships Dissolved" in the *Times*, January 19, 1842. The other partner in the firm was Daniel Price. He resigned in the late 1840s.

37. A British visitor met William Mackintosh at Culiacán in 1856. See C. Sevin, "A Journey to Mexico," *Journal of the Royal Geographical Society* 30 (1860): 45, www.jstor.org/stable/1799038.

38. The prospectus was published under the title *Compañía de Minas y rescate del mineral de Guadalupe y Calvo*, dated Mexico, May 19, 1836. It was reprinted in *El Noticioso de Chihuahua*, June 24, 1836.

39. Márquez Terrazas, *Pueblos Mineros*, 138 states that the El Zorrillo Company was formed in 1838, but he gives no source.

40. Sevin, "Journey to Mexico," 31.

41. Almada, *Guadalupe y Calvo*, 51.

42. Pell's letters are in the Samuel Bradford Birkbeck Papers held in the University of Sydney's Rare Books and Special Collections Library. I am grateful to Neil Boness, Sara Hilder, and Margaret Ashby for locating and sending me copies.

43. Birkbeck Papers, letters dated August 25 and September 9, 1837.

44. *Daily Commercial Bulletin* (Missouri), September 19, 1838.

45. A. M. Gilliam, *Travels in Mexico during the Years 1843 and 1844* (London: Wiley and Putnam, 1846), 326.

46. Stephen Douglas Buchan died from drowning in the River Tule and was buried in the cemetery at Guadalupe y Calvo on March 24, 1842: Guadalupe y Calvo parish registers, *Defunciones*, 1839–1860, images 72–73. The Guadalupe y Calvo parish registers are available online in the Church of All Saints Family History website. In references I have used the image number of the online film rather than the often illegible page number of the register.

47. Gilbert Pell to Samuel Bradford Birkbeck, September 9, 1837, Birkbeck Papers, University of Sydney. I assume the woman referred to in Pell's letter as "Mrs. Auld" was Mary Auld, who had married Robert Ogilvie Auld in 1834. It is not clear when or how she arrived at Guadalupe y Calvo.

48. The contract for this sale is in the Archivo General de Notarías, Mexico City (henceforth AGNM), Notary Ramón de la Cueva, acta 9635, año 1839, fol. 20617, dated August 10, 1839. It is signed by Manning and Mackintosh, acting as agents of the Guadalupe y Calvo Company.

49. The reports were produced for the eighth, ninth, and tenth annual general meetings of the company held in London on June 14, 1843, June 12, 1844, June 11, 1845. The pages that

survive consist entirely of the letters sent to the directors by Buchan and Auld. The reports are held in the New Mexico State University Library.

50. Gilliam, *Travels in Mexico*, 315.

51. R. A. Smith, *Borderlander: The Life of James Kirker, 1793–1852* (Norman: University of Oklahoma Press, 1999), 128–29.

52. Robert Auld's death was reported in the *Caledonian Mercury* (Edinburgh) on January 25, 1847.

53. Guanaceví was usually written as Guanasevi at this time. I have used the modern orthography.

54. Buchan's letters to Mackintosh are in the Manning and Mackintosh Papers. They are dated: March 3, April 23, April 30, and August 30, 1847.

55. This £6,000 may have been borrowed as an interest-free loan from the Anglo-Mexican Mining Company. In 1850, Christopher Richardson, a shareholder in the Mining Company, threatened to petition to have it wound up unless the directors took steps to recover the money from the Mexican and South American Company. The same people were directors of both companies and Richardson in effect accused them of corruption. See the letters in the *Mining Journal*, September 14, 1850.

56. M. G. Wigham to Manning and Mackintosh, March 1850, in Manning and Mackintosh Papers.

57. *English Law Reports*, 53, 867.

58. R. W. Randall, *Real del Monte: A British Mining Venture in Mexico* (Austin: University of Texas Press, 1972), 98–99, describes the Pachuca Company as a subsidiary of the Real del Monte Company because they had the same directors. Clearly, there were close links but shares in the Pachuca Company were always quoted separately and it seems to have operated independently. Randall (pp. 109–25) provides an excellent explanation of the various refining methods.

59. Another Pachuca Silver Mining Company was formed in 1860. See the *Daily News*, December 12, 1860 for the directors' report.

60. H. E. Cross, "The Mining Economy of Zacatecas" (unpublished PhD diss., University of California, 1976), 236.

61. Gennaro Placci, *Statement of Facts Relative to the Negotiation of Cerro del Bote, in the State of Zacatecas, Mexico* (London: A. Macintosh, 1838). It is dated Regent Street, [London], April 1838. I have found only a single copy of this pamphlet. It is in the possession of the bookseller, William Reese Company—see www.reeseco.com. I am very grateful to Mr. Reese for providing me with a copy.

62. Roger Allday may have been in London with Placci in 1838. He was certainly in Britain in 1842 and his first son, Roger, was born on December 16.

63. Cross, "The Mining Economy of Zacatecas," 236.

64. There are various letters from Kerrison and Co. in the Manning and Mackintosh Papers.

65. Roger Allday's movements are difficult to establish. He married Adelaide Thorp Stokes in New York on February 19, 1839, the day after her arrival from Britain. Their children were born in 1842, 1844, 1846, and 1849, all in Britain. It is not clear that the family lived in Zacatecas but it does seem certain that Roger Allday crossed the Atlantic more than once in the 1840s.

66. *Morning Chronicle*, December 6, 1849, mentions that the Bolaños Company "has recently been dissolved."

67. Cross, "The Mining Economy of Zacatecas," 239.

68. Sevin, "Journey to Mexico," 52.

69. *Ipswich Journal*, December 6, 1873 has the bankruptcy trustee's accounts of the sale of the Kerrison brothers' assets.

70. *Bury and Norwich Post*, December 9, 1873 has a full report of the creditors' meeting.

71. Edmund Roger Allday's unpublished autobiography is in the Kerrison family papers.

72. Among those who worked at the mine between 1860 and 1900 were Ernest F. Ayton, Alexander McGowan, and James Francis Ryan.

73. *Mexican Herald*, August 8, 1898.

74. Article in *St. Louis Globe-Democrat*, July 9, 1887.

75. *Birmingham Daily Post*, November 2, 1891.

76. These details of dividends are given in a letter from the London solicitors, Bailey, Shaw, and Gillett to Roger Allday Kerrison's solicitor, Arthur Preston, dated November 22, 1878. The letter is in the Birkbeck Papers. I am grateful to Margaret Ashby for locating the letter and to the librarian Sara Hilder for kindly providing me with a copy.

77. *Denver Evening Post*, December 1, 1896.

78. *Mexican Herald*, September 23, 1906.

79. *Fort Worth Star-Telegram*, October 5, 1906.

80. Roger Kerrison had been a director of Barclays Bank, Ipswich. He retired in December 1902.

81. The reports of the court case are in the *Times*, October 29, 1909 and April 16, 1910.

82. *Mexican Herald*, April 15 and November 25, 1908.

83. The Placci family left New York for Liverpool on the steamship Atlantic around May 26, 1856. *New York Times*, May 26, 1856.

84. I am very grateful to Nicholas Webb, archivist at Barclays Group Archives, for locating Roger Allday's bank account and sending me copies from it.

85. A son, Carlo Placci, was born in London on November 23, 1861. He became a very well-known writer.

86. *Ipswich Journal*, November 4, 1893.

87. *Times*, January 6, 1909.

88. I am grateful to Mr. Roger Kerrison for many of the details about the El Bote Company.

89. Birkbeck also had some small financial involvement in the Zacatecas Mining Company. See the accounts given in N. J. B. Plomley, *Several Generations* (Sydney: Wentworth, 1971), 134.

90. Birkbeck's journal is in two parts, covering the years from July 1829 to December 1854. It is a detailed record of his financial and business transactions: University of Sydney's Rare Books and Special Collections Library, Call. No. Birkbeck 7.3. I am very grateful to Neil Boness for providing me with a copy.

91. Plomley, *Several Generations*, 59, 135.

92. Ibid., 59.

93. Birkbeck and Damiana Valdez were married at Zacatecas on December 30, 1839. Damiana was probably aged fourteen or fifteen.

94. F. Palmer to Charles Birkbeck, August 30, 1878; correspondence in the Birkbeck Papers.

95. Plomley, *Several Generations*, 94.

96. A. Knight, "The Political Economy of Revolutionary Mexico," in *Latin America: Economic Imperialism and the State*, ed. C. Abel and C. M. Lewis (London: Athlone, 1985), 290.

Chapter Seven

Miscellaneous British Enterprises

THE MINTS

Silver and gold mining were always the principal areas of interest for British investors, but there were other closely related activities that also attracted their attention. The most productive mines were spread across the country, considerable distances apart and a long way from the capital, but the Spanish authorities had always insisted that all gold and silver must be transported to Mexico City for minting into coins. The only mint allowed in the colony, therefore, was the *Casa de Moneda* in Mexico City. It operated with much success for almost three hundred years, producing hundreds of millions of coins. The war of independence, however, brought the end to its monopoly as authorities far removed from the metropolis, for example, in Chihuahua, Durango, Zacatecas, and Guanajuato, found that it was too dangerous to transport silver to the capital, and as a result, they were running out of money. Hence, new mints were established in several of the mining regions during the war. They had only limited success because the amount of silver introduced by the miners quickly declined as mines were abandoned by their owners, fleeing the war zones. In the early 1820s, however, two new factors prevailed. First was the adoption of the federal system with eighteen autonomous states. Those new state authorities in the mining areas wanted their own mints from which they would derive taxation and other benefits. Second, with the arrival of the British with their money and machinery, it was assumed by everybody that the production of all precious metals was certain to expand and that modern and efficient mints would be essential. The state of Guanajuato, where much of the early British investment was introduced, was the first to decide that major investment in their mint was required.

On 31 May 1825, John William Williamson, the commissioner of the Anglo-Mexican Mining Company in Mexico, signed an agreement with the Guanajuato state government whereby a new company was to be set up to operate the state mint. The company was awarded the exclusive privilege of coining all the gold and silver presented to the mint for a period of ten years.[1] Once confirmation of this agreement reached London, the Anglo-Mexican Mint Association was announced and a prospectus published.[2] The bankers, auditors, and directors were much the same as the Anglo-Mexican Mining Company, and although always kept separate, the two businesses were to be closely connected in future years. Among the founding directors of the Mint Company were Robert Mushet and Edward Hurry. Mushet, described in the *DNB* as "an official of the Royal Mint," had worked for years in the Royal Mint and was clearly recruited for his experience and expertise in casting, refining, and other processes. Hurry was a shipping agent and insurance broker in partnership with John Diston Powles in several of his various companies at his business premises in Freeman's Court, Cornhill. The new company was to have a capital of £250,000, divided into 10,000 shares of £25, with the first installment of £5 due on 13 September. Investors were told that the amount coined at the mint before the war of independence had been 4,000,000 to 5,000,000 dollars a year but the present operators were so inefficient and expensive that miners were sending their bullion to Mexico City. Ample profits could, therefore, be expected from the more efficient methods that experienced British personnel would use.

Details of the company were in the newspapers early in September 1825 and the shareholders were reminded on the 7th that the Deed of Settlement had to be signed within twenty-one days at the office at 7 St. Helen's Place, Bishopsgate Street. Demand for the shares was not particularly strong, and although the opening price was £6, which was a premium of £1 on the first installment price of £5, the price soon fell back to around £5. Finally, it may be noted that at the first meeting of the directors on 1 November, George Canning Scott was appointed Master of the Mint at Guanajuato.

For the first few years, the directors of the new company concentrated on the reconstruction and re-equipping of the mint. Skilled engineers and other trades were recruited and sent to Guanajuato. Orders were placed at the Boulton factory at Soho, Birmingham, for coin presses, milling machines, dies, lathes, hand tools, and two steam engines. The first bill came to £8,281 and as more requests for machinery and men came in from Guanajuato, Mushet realized that it would be a long time before the company could make a profit. Williamson was instructed to negotiate an extension of the contract with the State authorities and the result was an extension to 1842. In the meantime, the machinery that had been ordered was manufactured at Soho. It took time, as did the training of more workers. Finally, it was finished and dispatched in October 1827. The directors decided not to send the steam engines, as they had been advised that there was not enough fuel—timber or coal—in the Guanajuato area to operate them. Two more of Soho's engineers—Edward Riley, a blacksmith, and Robert McCleish, a general mechanic—traveled to Mexico to supervise the use of the machinery. The newly refurbished mint building was opened on 14 November 1827 and production began.[3]

The Anglo-Mexican Mint Association retained the contract for the Guanajuato mint for almost fifty years. It was renewed in 1842 and 1853 but eventually in 1873, the then national Mexican government decided to terminate it. With no alternative options for the use of its capital, the directors and shareholders decided to wind up the company and a liquidator was appointed. Over that half-century, of course, the personnel both in London and in Guanajuato had changed regularly. Many more British managers, engineers, and mechanics went to Guanajuato, including John Potts, Edward Herry, Robert Traill, James Cottenell, John Graham, John Gemmill, William Flynn, John Willett, Edward Hoffay, John Stanley, and Lionel Brough. Some of these men whose contracts of employment were normally three years stayed in Guanajuato for many years, while others went to work for other British companies. Orders for dies and machinery continued to be supplied by the Boulton factory for many years; improvements were made in the manufacturing process; and new designs for the coins, both gold and silver, were commissioned.

The company's activities were not without controversy. In December 1841, represented by Manning and Marshall in the negotiations with the Mexican government, its contract had been renewed for a further fourteen years. A year later, in December 1842, again represented by Manning and Marshall, it was awarded the contract for the operation of the Zacatecas mint. In both cases, large sums of money had to be paid in advance to the government. The state authorities of Guanajuato and Zacatecas, aided by petitions from the clergy, merchants, local officials, and many other groups, protested against the contracts, waging a long and fervent campaign to have them revoked. Eventually, in 1845, after congressional reports and more heated debate and opposition, they were confirmed. British pressure through its diplomats in Mexico City was a major factor in the decision.[4] The renewal in 1853 was less controversial but even more expensive. An advance payment of 730,000 pesos was made as a loan to the government on the security of the annual rent of 90,000 to 100,000 pesos that the company agreed to pay for the exclusive operation of the mints at Guanajuato and Zacatecas. The contract was to run for ten years from 1856 to 1866.[5]

The company was able to pay large amounts in advance to win the contracts because control of the mints had proved to be a lucrative business. One estimate by an American newspaper claimed that the Guanajuato mint alone yielded a profit of 60,000 pesos a year.[6] Even more revealing perhaps was a comment in the London *Daily News* (4 April 1855). Its reporter had not been allowed into a recent meeting of the shareholders on the grounds that "the directors were afraid of exciting the jealousy of the Mexicans." Unfortunately, none of the annual accounts presented to shareholders has survived or been located, but the share price and dividends confirm that large profits were made.[7] In the early years, there was little movement in the share price but from 1830 onward, there was a steady increase as profits were announced and dividends were paid. In 1831, the price of a share on which £5 had been paid was £8. In 1843, on a £10 paid-up share, it was £12, and in 1851, £25. Throughout the 1850s to the closure of the company in 1873, the price on a £10 share remained between £15 and £20 and when the liquidator realized the assets, shareholders received two payments of £5. Few shares, it must be said, seem to have been traded, because there are many months in which there is no reference to the company in the stock exchange lists. There are no extant registers and hence no information on the identity of the shareholders. One exception was Roger Allday Kerrison, whose bank account shows that he received annual dividends of between £155 and £190. It is probable, however, that most of the shares were retained by the original directors and their families. They, and the other "proprietors" as they were called, received regular dividends annually, and occasionally, twice a year. The rate in 1834 was six shillings per £10 share. Throughout the 1840s, it was twelve shillings and in the 1850s, it rose to £1, paid twice a year. There is no doubt that the Anglo-Mexican Mint Company proved to be a sound investment and one of the very few British companies that had produced profits for its investors.

The second mint in which the British had an interest was at Culiacán, capital of the northern state of Sinaloa. The region was one of Mexico's important mining areas with numerous silver and gold mines but, with no local mint available, the miners had to meet the heavy cost of sending their metals to the distant Durango or Guanajuato mints. There was clearly a good case for a mint in the state. On 11 April 1832, from the small town of El Rosario on the Pacific coast of the state, an employee of the London merchants, Finlay, Hodgson, and Co., wrote to his employers. Unfortunately, he failed to sign the letter and we do not know his name.[8] He wrote that the state authorities had decided to establish a mint. Although he does not say so directly, he seems to have been commissioned to make the necessary arrangements. He explained in detail what machinery was required: cutting out presses, turning lathes, tools, milling machines, dye presses, and a rolling mill to be operated by mules. Fully trained British engineers, blacksmiths, carpenters, and other trades would have to be found and contracted to set up and operate the new mint.

Finlay, Hodgson, and Co. accepted the proposal of their anonymous employee and in July 1832, they contacted the Boulton factory at Soho, asking for an estimate for manufacturing the machinery that had been requested.

The cost was calculated to be £1,375—the final cost was £1,731—which Finlay, Hodgson agreed to pay. Two employees were selected for the job of establishing the mint. They were Jacob Peat and Joseph Griffiths. Peat was an engineer and he was given a five-year contract on a salary of £200 to £300 a year plus travel and medical expenses. Griffiths was a blacksmith and his salary was set at £150 to £200 a year. Both spent several months at Soho training in the use of the minting equipment. They set out from Birmingham in February 1833 on the long journey to Culiacán, traveling across Panama to the Pacific and up the coast to Sinaloa. Griffiths died not long after arriving in Culiacán in the summer of 1833, possibly a victim of the cholera epidemic of that year, and it was left to Peat to begin the construction and equipping of the mint.[9] The machinery, dispatched from Birmingham in October 1832, arrived at Mazatlán, from where it was gradually transported to Culiacán. Progress was slow, not just in the construction schedule, but just as important, in finding entrepreneurs willing to take on the job of managing and operating the mint. No offers were forthcoming and on 10 January 1835, the State government issued a formal invitation or *convocatoria* to businessmen to take over the mint.[10] It advised that the mint building was largely finished and that the machinery was in storage in Culiacán and Mazatlán. Proposals to operate the mint were invited.

From the little information available, it seems that nobody wanted the Culiacán mint, and all the equipment and tools manufactured in Birmingham were left in storage. There are several unanswered questions. We do not know, for example, what the intentions of Finlay, Hodgson and Co. were in the first place when they agreed to participate in the creation of the mint in response to the letter from their unnamed employee. We have to assume that further correspondence took place, that perhaps contracts were signed locally, and that they were reimbursed by the State for the £1,731 they spent at the Soho factory. The lack of interest by local or Mexico City-based individuals or companies in managing the mint is also surprising. It may have been the result, as Gutiérrez suggests, of the very unstable political situation in 1835, when the process of dismantling the federal system in favor of a centralized republic was being initiated. The autonomous states like Sinaloa were to be abolished under the new regime and replaced by departments ruled from the center in Mexico City.[11]

No progress was made, therefore, with the Culiacán mint for several more years and as far as we know, British interest in it lapsed. Then, in 1843, a detailed proposal to open a mint with an attached *Apartado* or separation plant, was put to the national government. The person whose name was attached to the document was José Delmotte, and he was acting on behalf of the Guadalupe y Calvo Mining Company. The proposal was accepted and the Guadalupe y Calvo Mining Company was awarded the contract to operate the mint for fourteen years during which it had the exclusive right to produce gold and silver coins within the department of Sinaloa. Work to complete the building and install the machinery that had been bought a decade before was slow in starting and it was not until 1846 that the first coins were produced. The cost of the installation was said to be 350,000 pesos, but where the money came from is not known, although it is clear that Ewen Mackintosh once again became a key figure.[12] He, it will be recalled, was the agent and treasurer of the Guadalupe y Calvo Mining Company. The capital for the mint may have been raised from its shareholders but there are no records to confirm this supposition. What is certain is that Mackintosh had ten shares valued at 40,000 pesos.[13] Whether he was holding those shares on his own account, or on behalf of other, possibly British investors, is not known. Finally, while British capital may not have been invested, British labor and expertise was certainly to be found in the Culiacán mint. Among the Britons employed there in the late 1840s and early 1850s were Louis Price, George le Brun, Thomas Millen, Quentin Douglas and Ewen Mackintosh's nephew, William.

The next mint to interest British investors was in Chihuahua. There had been a mint in this northern state since 1811, but it had operated only intermittently and without producing much in the way of profit for its operators. From 1832 to 1839, it was under the direct control of state-appointed managers, but in March 1839, terms were agreed with John Kelly and his partners, John Jennison and John Potts, to take over the running of the mint. John Kelly was British and he had been in Mexico for several years working as a merchant, mostly at Mazatlán. His business had prospered and enabled him to spread his investments into mines and other local enterprises. His contract for the mint at Chihuahua, agreed on 7 March, was approved by the national government on 11 July. Not long afterward, it seems, John Potts left for Britain to buy all the tools and machinery needed to modernize the mint. He was an experienced engineer who had been working for the Anglo-Mexican Mint Association at Guanajuato since at least 1834. Doty's research found him at the Boulton factory in Birmingham in early 1840, where he requested an estimate "of the expense of original and moneying dies of the doubloon [onza], dollar, real, and half reals, similar with a slight variation to those made for the Anglo Mexn Mint Asson."[14] Following various technical adjustments, the order—twenty-four pairs of dollar dies and six pairs each of real, half-real, and doubloon dies packed in iron tubes—was completed by the factory and dispatched to Liverpool in March 1840. Potts may have sailed back to Mexico on the same ship to keep an eye on his purchases, and on his safe return to Chihuahua he began the manufacturing process, producing coins from the new dies. He stayed in Chihuahua, working for the mint, for several more years, placing further orders with the Soho factory in 1848. By that time, he had been joined by his brother, Orlando Frederick Potts, and several other British workers including another Potts, Francis Edmund, who was also possibly a relative, and George Parker.

There is no information available on the amount of money that was invested in the Chihuahua mint by John Kelly and his partners, nor do we know where the money came from. It may have been from Kelly's own resources or there may have been British-based investors, possibly recruited by Potts during his time in Birmingham. We know that Potts remained in Mexico until at least the 1860s. He invested in mines in Chihuahua, San Luis Potosí, Pachuca, El Chico, and elsewhere. Although details are scarce, it seems that he also retained his interests in the Chihuahua mint. In 1863, a new company was announced in London to invest in Mexican mines. This was The El Chico Silver Mining and Reduction Company and among its directors listed in the company prospectus was "John Potts, Esq. Mexico, Mint Owner."[15] Seven years later the Guerrero Gold Mining Company was announced in London and its Managing Director in Mexico was John Potts.[16]

The success of the Guadalupe y Calvo mines soon brought about projects for a mint in the expanding new town. Once again, it was the British Mexican and South American Company that put forward the first project. Some time around June 1839, Manning and Marshall, acting as agent for the London company, submitted a proposal for a mint. It was referred to congress on 22 June and a Treasury committee report on the idea was discussed in the chamber of deputies on 3 July.[17] There was some strong opposition led by deputy Pedro Barajas. He did not agree with the privilege of coining money being given to a private company or individuals, and although he did not spell it out, to foreigners. His opposition was enough to persuade the committee to withdraw its report on 4 July.

The failure to get congressional approval for a mint financed and managed by the British Mexican and South American Company seems to have persuaded Manning and Marshall, or Ewen Mackintosh, to try another strategy. Just a week after the congressional debate, they appeared before a notary in Mexico City. Acting on behalf of the Mexican and South American Company and the Guadalupe y Calvo Mining Company, they transferred to José Delmotte full authority "to use whatever measures, without exception, he felt necessary" to persuade congress to grant a licence to open a mint at Guadalupe y Calvo.[18] In other words, the request was to be from the Mexican-based as well as the London-based company.[19] A new proposal was introduced in congress in late July or early August 1839 and debate on a Treasury committee report that recommended approval began on 7 August. Thereafter throughout August and September the deputies examined each article in the report line by line, approving some, rejecting others, or referring them back for revision. Hours of debate were devoted to the subject and eventually approval was granted.[20] The proposal was sent to the senate, and it gave its approval on 4 December, ending its discussion with the words "al gobierno" ("to the government")."[21] The Executive was instructed to arrange with the Mexican and South American Company and the Guadalupe y Calvo Mining Company the establishment of a mint at Guadalupe y Calvo. The contract for the operation of the mint was for ten years, starting from the day when the first coins were produced. They were to be the same as those produced by the Mexico City mint that would provide all the necessary dies. The companies were to bear all start-up and operational costs.

The British Mexican and South American Company, therefore, acting with the Mining Company, won the contract for the mint at Guadalupe y Calvo in December 1839. For reasons that are unknown, however, the Mexican government failed to implement the congressional decision. Approval was eventually given in the autumn of 1841 but not acted on and it was not until 3 October 1842, after more discussion by government departments, that a law was promulgated giving the concession to Delmotte, representing the Guadalupe y Calvo Mining Company. There is no mention of the Mexican and South American Company in the decree and it is not clear if it was still involved. A week later, Delmotte set up a company to build and manage the mint.[22] A large amount of machinery and other equipment was soon ordered from France, and construction of a building to house the mint was started. Minting finally began on 1 June 1844 and over the next six years, 2,063,958 silver and 2,311,104 gold coins were minted.[23] We do not know who paid for the mint and in particular, if there was any British investment. The capital invested appears to have amounted to 557,000 pesos and of this, 220,000 pesos was attributable to Manning and Mackintosh.[24] It held shares valued at 120,000 pesos on its own account and the other 100,000 may have come from the Mexican and South American Company.[25] It may be that other Guadalupe y Calvo Mining Company shareholders, including Buchan, Auld, and Henry Mackintosh, put up some money, but we have no details of their involvement. As in the other mints around the country, however, the British were very much involved in the workforce. The first director of the mint was Thomas Mackintosh and the final director when the mint closed in 1851 was Matthew Wigham. Other British employees included James Scobie, Richard Killaly, and John Brown. A balance sheet dated 30 June 1850 has the names of Henry Mackintosh, Thomas Lambie, Quentin Douglas, Louis Price, John MacDonald, Edward Clark, Randell and Co., and Bowring and Co. but it is not clear what their connection was.[26]

Finally, on the issue of mints, it is worth noting that Ewen Mackintosh also held the contract to manage the Mexico City mint from 1847 to 1849 but there is no suggestion or evidence of any British capital being involved. It has been calculated that at that time, Mackintosh was in control of more than 80 percent of all silver minted and almost 90 percent of the gold.[27] What is by no means clear, however, is to what extent he was acting on his own account, or on behalf of investors, including the British.

OTHER VENTURES

Silver and gold mining with the associated business of coining money were clearly the main areas of the Mexican economy to attract the attention of British investors and British personnel. As far we know, and further research may reveal others, the Pachuca Mining Company was the last joint stock company formed in London in this second phase of British investment in Mexico's mining industry from 1830 to 1860. As we have seen, the El Bote Mining Company tried but failed to attract investors in London in 1850 and thereafter was operated as a privately owned business. There was, however, significant other British interest and investment in mining companies formed not in London but in Mexico and it is possible—again more research is needed—that some of the capital used in acquiring shares in these new ventures originated in Britain. British miners and merchants working in Mexico were always on the lookout for promising old or new mines that came on the market and they seem to have had no trouble in raising the considerable sums needed to bring the mines into operation. Unexpected encouragement came from the Mexican government in 1842. On 12 July, a law was issued whereby for the first time foreigners were allowed to own mines, provided they could demonstrate that they were restoring old or abandoned properties. Hitherto, foreigners had only been allowed to acquire shares in mines with the ownership always retained by Mexicans. Two other laws in the same year were also relevant. The first was dated 11 March. It stipulated that foreigners who were out of the country for two years or more lost any rights they had to mining properties. In October, this law was amended and absentee foreigners were allowed to retain their mining rights, provided the companies in which they were shareholders were still in existence. This amendment was issued in response to an appeal made by a British miner, William Henry Egerton.

Egerton had arrived in Mexico in 1829 to work for the British mining companies. He was followed a year later by his brother, Daniel Thomas Egerton. Daniel was an accomplished artist and after touring the country and making sketches of the scenes he encountered, in 1835 he returned to Britain, where he exhibited several of his Mexican paintings to great acclaim. In 1842 he returned to Mexico City, accompanied by Agnes Edwards. They rented a house in Tacubaya, then a village about three miles from the city center. In what became a notorious case, the couple were brutally murdered in the garden of their property on 27 April.

We do not know the whereabouts of William Henry Egerton when his brother was murdered, but it is possible that he was in El Oro, a mining district in the state of Mexico near the border with Michoacán. Several of the mines in the area had been worked by British companies in previous years but then abandoned. Almost certainly in response to the law allowing foreigners to have outright ownership of foreign mines, Egerton and several associates formed a new company entitled the *Compañía Restauradora del Mineral del Oro*. The deed establishing the company was drawn up by a notary public in Mexico City on 30 September 1842. The main shareholders were Francisco Murphy, with twenty shares, Egerton with eleven, and Byrius Hooten and Company, a British merchant partnership based in Mexico City, with ten. Murphy was a wealthy mine owner, entrepreneur, and speculator with very close connections to the Mexican government. Originally Spanish, he was by this time a naturalized Englishman. His brother, Tomás, was Mexico's chargé d'affaires in London. There were three other shareholders with seven shares between them. They were José Rafael Oropeza, Lorenzo Carrera, and José Marzán. The three Britons were, therefore, the majority shareholders and as such had to provide most of the working capital that was set at an initial 80,000 pesos.[28]

Egerton continued to work for the company until at least 1847 and by that time there were more British miners employed, for example, Robert Glennie. Other Britons were also involved as partners and shareholders in later years. They included William and Sarah Jamison, Andrew Melville, Richard Dillon, George and Charles Whitehead, Alexander Grant, George Murray, Nathaniel and Lionel Davidson, and finally, the ever-present Ewen Mackintosh. Their investments in the company did not produce the hoped-for dividends. In 1854, it was stated that "more than 2,000,000 pesos had been expended on the mines with no return except the ruin of the majority of the investors."[29] The mines of El Oro had, in the words of Dr. Staples, brought disillusion and grief to one owner after another.

The Del Oro Company was not the only one in which British investors were represented. On 14 February 1846, the *Compañía de Minas denominada Restauradora del Mineral de Catorce* was formed. The capital was 200,000 pesos in twenty shares of 10,000 pesos, each divisible if required into quarters at 2,500 pesos each. Among the promoters and shareholders were John Richard Rule (two shares); Ewen Mackintosh (two shares); Henry Mackintosh (one half share); Edwin Holdsworth (one half share).[30] John Rule, an experienced Cornish miner, and several of his brothers, had worked for twenty years for the British Real del Monte Company, based at Pachuca. He was appointed director or manager of the new Catorce company. As we have seen earlier, Ewen Mackintosh was the leading British merchant in Mexico and his brother Henry was an experienced miner who had been working for the Guadalupe y Calvo Company in Chihuahua for several years. Edwin Holdsworth was a merchant in Mexico City, again having been in the country with his large family for many years.[31] Few details are available of the future operations of the company at Catorce, but we do know that in March 1850, a party of ten Cornish miners recruited by Rule arrived at

Tampico. A few weeks later, a cargo of "iron and mining implements" also arrived for the Catorce Company.[32]

Another similar company that attracted British investors was the *Compañía Restauradora del Mineral del Doctor*. The mining district known as *El Doctor* was in the state of Querétaro, and it had also been worked earlier by several groups, including British companies, only to be abandoned as unlikely to yield a profit. Little is known about the new company except that it was formed in January 1848 by a group of British investors. The Cornishman William Rule was again involved and his partners included William Higson and Benjamin Barton. Both were Mexico City-based merchants who had been in the country for several years. Higson had arrived in 1827 to work for O'Ryan and Co. in Mexico City. Subsequently, he seems to have been in the clock and jewelery trade and in later years, he worked as a general merchant. Barton was a partner in Bates, Barton and Company. Three other Britons were shareholders: William Jamison, Robert Allen, and John Frener. Although the details are far from clear and individual holdings are not known, the capital appears to have been 42,000 pesos, divided into twenty-one shares of 2,000 pesos.[33]

The two *Compañías Restauradoras* that we have described so far were by no means the only ones to be set up in the 1840s and early 1850s. Similar ventures were organized to work mines in other states and British investors were among the shareholders. Where or how the money was found to buy into these always-risky mining companies is not known. Some of the individuals concerned, for example, William Rule, John Buchan, Robert Auld, or James Nelson Shoolbred, had all worked for the earlier and much larger British companies such as the United Mexican Company, the Real del Monte Company, or the Anglo-Mexican Company. With very few exceptions, these British companies had lost substantial amounts of money in their operations, but it seems that their senior managers had personally done very well financially. They had accumulated enough funds to be able to speculate and invest in mines and other businesses on their own account. The Rule family members in particular had widespread investments in mines in Querétaro, Pachuca, Catorce, and elsewhere. Buchan and Auld were likewise not short of money. In 1846, they agreed to go into partnership with James Randell to found Randell and Company of Durango. Buchan contributed 24,000 pesos and Auld 12,000, with Randell paying 26,000 so that the firm had a working capital of 62,000 pesos. Randell was to be the managing and active partner and Buchan and Auld "dormant partners" taking no part in the daily business. Some British merchants, mainly making their profits from imports and exports and in some cases, financial speculation in government debt bonds, were also willing to risk their capital in mining ventures. Benjamin Barton, for example, owned shares in Zacatecas mines and Bates, Stokes, and Company had invested in the Tlalpujahua mining district. George Best, a merchant apparently of German origin but with a British passport, was a shareholder in a company set up in 1851 to exploit mines in Taxco, and he also had shares in several of the Pachuca mines.

How profitable these ventures were is unknown but in one instance, a mining investment proved to be a genuine bonanza. This was the well-known case of the Almaden mercury mines, discovered in 1845 in California, then a Mexican territory. Andrés Castillero, a Mexican army officer who had gained possession of the cinnabar deposits, was unable to raise enough capital to work them. He sought investors and among those who saw the potential were two British merchants, Eustace Barron and Alexander Forbes. They had developed a highly successful commercial company at Tepic on the Pacific coast, and had invested widely in other areas, including mining. They poured several hundred thousand dollars into the Almaden mine and within a few years were deriving very large profits from it. Even after California became part of the United States following the war with Mexico, the British company retained control of the Almaden mine.

Whether these and similar investments all came from personal funds or capital raised in Britain is not known. We have seen how London companies raised capital by selling shares on the Stock Exchange but it is also probable that some of the London-based merchant bankers and financiers used their own money for private rather than public investments in Mexico. John Schneider and Company, for example, were, as we have seen, major investors in the mining companies and in the Mexican and South American Company. They were also very active in the bond market and in 1845 they were awarded the lucrative contract to act as the Mexican government's official agents in Britain and Europe.[34] Shortly afterward, in partnership with Manning and Mackintosh, they became involved with a controversial scheme to create a canal across the Isthmus of Tehuantepec. The validity of their contracts for this project that also involved the colonization of adjacent lands was challenged in the courts and their interest was soon relinquished. C. de Murrieta and Company was another prominent London business which may have had private investments in Mexico. The Mexican and South American Company decided to liquidate its Mexican assets in 1847, but it appears that it retained some interests in mines. On 22 November 1848, John Buchan, still the company's representative in Mexico, transferred one of its mining interests in Guadalupe y Calvo to Pedro Romero.[35] Finally, there was N. M. Rothschild and Sons. The Rothschilds were of paramount importance to the Mexican mining industry because they largely controlled the world supply of mercury, an essential ingredient of the amalgamation process. For the first few years after independence, they conducted their commercial business in Mexico through William Drusina and Company, a British-owned merchant company based in Mexico City. Then in 1843, they decided to send a cousin, Lionel Davidson, to develop their interests. He was soon joined by his brother, Nathaniel Davidson. The two Davidsons began to seek out commercial opportunities and were soon investing heavily in the mining industry, buying shares in mines in Guanajuato, Temascaltepec, Taxco, Zacatecas, and elsewhere. Some of these investments seem to have been on their own private account while others were for Rothschilds.

While it is clear that mining was the major attraction, it was not the only area of the Mexican economy that interested British investors. Nathaniel Davidson became the owner of the biggest iron foundry in the country, costing 87,771 pesos and providing work for 113 workers and employees.[36] Richard Francis was a partner or shareholder in several mines and he had substantial investments in iron and nitrate works. He was also the owner of the *tienda de raya* or estate shop at Real del Monte.[37] There was significant British interest in the textile industry. In 1830, anxious to reduce the country's dependence on imports, especially British textiles, the Mexican government founded a credit bank, the *Banco de Avío*. It was to be provided with 1,000,000 pesos, drawn from customs revenues, to be employed in promoting industrial development, especially textiles. Over the next few years, loans from the Bank were used to start textile factories. Then, in 1837, to help this new domestic industry, the import of coarse cotton cloth and yarn was prohibited. A rapid expansion of textile factories followed: "From four cotton mills in 1837, the industry expanded in less than a decade to more than forty mills—representing a 10 million dollar investment."[38] Some of the British merchants who had been making large profits from their imports of British textiles objected strongly and loudly to the restrictions but others saw an investment opportunity. They began to set up their own textile factories and workshops. Among the newcomers were Washington West, John Biggs, and Thomas Grandison, who set up the Mexican Industrial Company to operate a new factory producing cotton textiles. Other leading figures were the Hope family led by two brothers, Archibald and Cuthbert. They had arrived in Mexico in the late 1820s and operated as general merchants, directing much of their business to a branch in Liverpool managed by other family members. In March 1839, they began to invest in textiles. First, together with a Mexican partner, they bought a haberdashery in Mexico City, investing 29,750 pesos. A few weeks later, they rented property in which they set up a textile factory producing yarn and cloth. In later years, they established several other workshops in the capital and they bought shares in textile factories in Jalapa, Guadalara, Tlalnepantla, and elsewhere. On the west coast at Tepic, Eustace Barron and Alexander Forbes operated cotton mills that had as many as 5,000 spindles working day and night, producing very large profits for the partners.[39]

Another British investor in the textile industry was George Ainslie. He arrived in Mexico City in 1828 where he was soon joined by his brother, Alexander. Both were master plumbers, and it was not long before they had built a profitable plumbing business, winning several contracts from the metropolitan council to install water pipes in public buildings. They invested in lead mines and opened their own factory in Coliseo Street to manufacture pipes. Qualified plumbers were recruited from Britain, for example, Ronald McDonald and David McCulloch, and Mexican apprentices were trained. In 1838, the Ainslies made their first entry into textiles when they rented a property they owned to John and Walter Bailey, who wanted to start a factory producing shawls (*mantas tejidas*). The Ainslies provided sixty looms. The following year, with a loan of 20,000 pesos from the *Banco de Avío*, they converted a hand-weaving shop in Mexico City to power spinning and weaving and in later years, a *Fábrica de Blanquear y Estampado de minas* for the manufacture of white cloth (*piezas de mantas blanqueadas)*.[40] They also had shares in other textile factories. Finally, it seems that they did not always get paid for their plumbing work. In a list of claims against Mexico compiled in 1862 and recognized by the government as correct, George Ainslie is owed 55,172 pesos for "contracts for plumber's work."[41]

Other British merchants who owned or had investments in textile factories were Andrew Lyall, John Corrigan, Joseph Welsh, and William Earl.[42] Together with Thomas Grandison, Archibald Hope, and George Ainslie, they signed a petition in February 1841 to the British Minister Plenipotenciary, Richard Pakenham. They explained that in response to the 1837 Tariff that had banned the import of cotton yarn and common cotton cloth, "we have embarked large capitals in the establishment of works for the manufacture of these articles."[43] In direct contravention of its own tariff, the Mexican government had recently decided to authorize the import of cotton products. They demanded that Pakenham should protest against this decision. He refused, saying in blunt terms that the British textile manufacturers in Mexico were in competition with those in Britain and that the British government's policy was free trade. Hence, he welcomed the new imports that would largely come from Britain. Hope, Ainslie, and their colleagues would have to put up with the competition. Further angry exchanges ensued. Pakenham did not change his mind but fortunately from the viewpoint of the petitioners, the Mexican authorities decided to revoke the licenses to import cotton goods that it had earlier approved.

It was not just the British merchants who were prominent in the textile industry. Skilled workers and equipment were also brought from Britain. Jacob Robertson, a Scot, managed the Miraflores factory near the capital. Thomas Grandison, another Scot from Glasgow, was the director of a large cotton factory at Cocolapan, Orizaba, and among his employees were "12 to 18 foreigners (mostly Scotchmen)." Almost all the machinery used in the factory was supplied by the Glasgow firm of James Grandison and Sons. It had also "sent over to them a farmer from Ireland, acquainted with the cultivation of flax, and a supply of seed at the same time."[44] In 1835, the

Mexican entrepreneur and industrialist, Esteban de Antuñano, contracted ten British textile engineers for his recently inaugurated textile factory in Puebla, *La Constancia Mexicana*. Apparently they had to travel incognito for fear of violence against them by local weavers whose livelihoods were threatened by the mechanization of their industry.[45]

Lack of research means that we have scant information, but it is certain that British companies and many British workers were also engaged in the cochineal trade.[46] Henry Ward noted that English merchants were involved in the trade as early as 1758, and by 1828, Britain's annual import of Mexican cochineal had risen to 139,586 lbs. Although the amount declined for a short time—95,728 lbs. in 1831—by 1840 it had increased to 490,000 lbs.[47] Manning and Mackintosh were said by one observer to ship regularly "five or six hundred bales, of 200 pounds weight each" from their warehouse in Mexico City. The cargo was just as valuable as the "treasure" that was also loaded on their carts.[48] To a limited extent, there were also British investments in land with successful merchants buying rural estates. The Barton family were part owners of a springwater business that they bought from another Englishman, Adam Turnbull. In the far south, other British entrepreneurs developed timber-exporting businesses, exploiting the vast range of mahogany and dyewoods to be found in the regions bordering on Central America. Edward Penny and William Bates were among several British shareholders in the company set up by the government in 1830 to manage the tobacco industry. Many other British merchants around the country in Guanajuato, San Luis Potosí, Zacatecas, Durango, Tampico, Veracruz, and Mazatlán invested their money in local businesses, property, mines, and factories.[49] One industry with a noticeable British presence was the manufacture of paper. Several factories or paper mills were opened within the metropolitan area and some were owned or managed by Britons. For example, there was William Benfield. He had arrived in Mexico to work as a tailor but soon acquired enough capital to invest in paper factories. Government publications began to appear carrying the words, "Impreso en papel megicano, de la fábrica de los Sres. Benfield y Marshall, calle de la Palma, núm 4. 1844." William Earl was a jeweler and clock repairer but he too in 1837 had enough money to open a paper factory at San Angel, then a village on the outskirts of the capital. James Arthur, Henry Gibbon, and Alexander Marshall were three other Britons who owned or managed or had shares in paper factories. These examples of British entrepreneurs are no more than a sample. Many other instances could be cited of British managers and skilled workers producing porcelain, furniture, carriages, beer, and a myriad of other products.[50]

It is evident that British investment in Mexico certainly did not end as a result of the events of the 1820s. The stock market collapse and its impact on the British mining companies; Mexico's default in 1827 on its foreign debt owed to British investors; the country's collapse from constitutional rule into widespread domestic political conflict; the failure of the British mining companies to make profits and pay dividends; the hostile environment; all were good reason to suppose that British merchants and capitalists together with ordinary investors who liked to play the market would choose to put their money elsewhere. Yet, as we have seen, more mining companies were successfully floated on the Stock Exchange after 1830—the Peñoles Gold Mining Association, The Zacatecas Mining Company, The Mexican and South American Company, and the Pachuca Mining Company. British capital was invested in mints at Guanajuato, Zacatecas, Culiacán, Chihuahua, and Guadalupe y Calvo. British entrepreneurs invested in textile factories, paper manufacture, cochineal production, timber exports, and many other industrial activities. London, Liverpool, Manchester, Glasgow, and Dublin-based merchants were not just prominent in Mexico City but were also active in all the main cities and ports of the country. Furthermore, Britons of all trades from carpenters to accountants, and plumbers to doctors, could be found everywhere. Finally, among the more unusual occupations for a British citizen was a Mexico City policeman. Anthony Gahagan joined the police at a salary of eighteen pesos a month and he was given a uniform. Around October 1848, he was arrested on suspicion of having joined a group of rebels plotting to overthrow the government. He was, he alleged, forced to sign a confession written in Spanish. Appealing for help to the British legation, he claimed to have no money or food and was destitute. Diplomatic intervention seems to have secured his release and Gahagan promptly obtained work with a mining company at Catorce.[51]

NOTES

1. There is a copy of the contract in NA, Mint 13/200, fols. 1–5.
2. There is a copy of the prospectus in BL, 1890.c.6.
3. These details are from R. Doty, *The Soho Mint and the Industrialization of Money* (London: British Numismatic Society, 1998), 232–46.
4. It is interesting to note that as early as 1843, the chairman of the company, John Schneider, wrote to the Foreign Office to ask for assurances that if the contracts for the Mints were challenged, he could count on British government support. He was told that the British government did not get involved in any way with private contracts: NA, F.O. 50/170, fols. 19–20.
5. Details of the 1853 contract are in the *Times*, December 21, 1853.
6. *Daily National Intelligencer* (Washington, D.C.), December 8, 1846.
7. One reason why there is so little information available about the company is that the directors—for reasons unknown—fell out with the editors of the *Mining Journal*, whose reporters were barred from the public meetings of the shareholders: see *Mining Journal* (1842): 14
8. The letter and others papers relating to the Culiacán mint are in the Boulton Archive (Birmingham Central Library), MS 3782/3.206.
9. Peat and Griffiths had arrived in Mexico by May 1833. Both requested security passes on the 23rd of that month. They were described as "*artistas*" who were to live in El Rosario.

10. *Convocatoria, que el supremo gobierno del estado de Sinaloa hace para empresarios del establecimiento de Casa de Moneda* [Culiacán, 1835].

11. B. Gutiérrez, *Memoria sobre la creación y estado actual de las Casas de Moneda de la República* (Mexico: Tipografía de M. Murguía, 1849), 21–22.

12. C. Sevin, "A Journey to Mexico," *Journal of the Royal Geographical Society* 30 (1860): 11.

13. AHNM, Acta 38851; notary Francisco de Madariaga; fol.1424; August 20, 1850.

14. Doty, *The Soho Mint*, 253.

15. The prospectus is printed in the *Times*, December 12, 1863. Potts may have been expelled from Mexico in 1859 but we do not know the cause. He subsequently claimed compensation of twenty thousand pesos for "indemnity for expulsion" and his claim was among those supported by the British government.

16. Prospectus in *Trewman's Exeter Flying Post*, April 27, 1870.

17. *Diario del Gobierno*, July 10, 1839.

18. AHNM, Acta 9592; notary Ramón de la Cueva; fol. 20575; July 10, 1839.

19. *Mining Journal* (February 22, 1840) reported that the Mexican and South American Company "are about getting a privilege to establish a mint in that district [Guadalupe y Calvo]."

20. The congressional debates are reported at length in the *Diario del Gobierno* throughout August to November 1839.

21. Senate session reported in *Diario del Gobierno*, December 10, 1839.

22. AHNM, Acta 19969; notary Ramón Villalobos; fol. 31725; December 13, 1842.

23. F. R. Almada, *Guadalupe y Calvo* (Chihuahua, Mexico: [s.n.], 1940), 57–60.

24. Undated letter from M. Wigham to Ewen Mackintosh, in Manning and Mackintosh Papers.

25. AHNM, Acta 38851; notary Francisco de Madariaga; fol. 1424; August 20, 1850.

26. Manning and Mackintosh Papers.

27. C. Avila Velasco et al., *Estado y Minería en México (1767–1910)* (Mexico: Fondo de Cultura Económica, 1988), 143–49.

28. AHNM, Acta 19080; notary Francisco de Madariaga; fol. 30832; September 30, 1842.

29. These details are from A. Staples, *Bonanzas y Borrascas. Mineras. El Estado de México, 1821–1876* (Mexico: El Colegio Mexiquense, 1994), 134–40.

30. *Plan de reorganización de la Compañía de Minas denominada Restauradora de Mineral de Catorce* (Mexico: Ignacio Cumplido, 1851).

31. The deed forming the company, dated February 14, 1846, is within the *Plan de reorganización*, 131–43. Henry Mackintosh replaced Rule as the director in 1848.

32. Letters from Stewart, Jolly, and Co. at Tampico, to Manning and Mackintosh, dated March 28 and June 4, 1850, in Manning and Mackintosh Papers.

33. AHNM, Acta 32554; notary Ramón de la Cueva; fol. 199; 31/01/1848.

34. M. P. Costeloe, *Bonds and Bondholders: British Investors and Mexico's Foreign Debt, 1824–1888* (Westport, CT: Praeger, 2003), 59.

35. AHNM, acta 34654, Notario Joaquín Vigueras, fol. 2278, 22/11/1848.

36. J. Bazant, *Alienation of Church Wealth in Mexico* (Cambridge: Cambridge University Press, 1971), 85.

37. AHNM, acta 65678; notary Francisco Calapiz y Aguilar, fol. 155030, 26/01/29.

38. D. W. Walker, *Kinship, Business, and Politics: The Martínez del Río family in Mexico, 1823–1867* (Austin: University of Texas Press, 1986), 138.

39. Report in the *Daily News*, January 23, 1850.

40. R. Potash, *Mexican Government and Industrial Development in the Early Republic: The Banco de Avío* (Amherst: University of Massachusetts Press, 1983), 209n59.

41. "Despatches Relating to British Claims," *Parliamentary Papers*, 64 (1862): 85–98.

42. There are useful lists of factory owners included in the "Memorias sobre el estado de la agricultura e industria de la república" (1843, 1844, 1845) published in *Obras de D. Lucas Alamán. Documentos diversos*, vol. 2 (Mexico: Editorial Jus).

43. NA, F.O. 50/144, fol. 115–23. Potash, *Banco de Avío*, 151–52.

44. See the appendix in W. P. Robertson, *A Visit to Mexico* (London: Simpkin, Marshall, and Co., 1853), 476–78.

45. W. L. Bernecker, *De Agiotistas y Empresarios. En torno de la temprana industrialización mexicana (siglo XIX)* (Mexico: Universidad Iberoamericana, 1992), 118.

46. The British consul at Veracruz noted in 1838 that there were many Britons engaged in the cochineal trade: NA, FO 50/119, fol. 200.

47. Import statistics from J. R. McCulloch, *A Dictionary, Practical, Theoretical, and Historical, of Commerce and Commercial Navigation* (London: Longman, Rees, Orme, Brown, and Green, 1832), 283; (1837), 301. W. Waterston, *A Cyclopaedia of Commerce, Mercantile Law, Finance, Commercial Geography and Navigation* (London: Henry G. Bohn, 1844), 164; H. G. Ward, *Mexico in 1827*, 2 vols. (London: Henry Colburn, 1828: 2nd expanded ed., 2 vols., London: Henry Colburn, 1829), 1:83.

48. Robertson, *A Visit to Mexico*, 249.

49. One of the most important that I have not mentioned but whose activities have been superbly described by Dr. Walker are the Martínez del Río family. Originally Panamanian, they became naturalized British subjects. See Walker, *Kinship, Business, and Politics.*

50. One of the best sources for Britons involved in Mexican industry is Bernecker, *De Agiotistas y Empresarios*. He cites numerous examples from many parts of the country.

51. The papers concerning Gahagan's case are in NA, FO 50/246, fols. 119–35.

Chapter Eight

The Shareholders

> It is foolish to be sober
> When all the nation's drunk. [1]

With a couple of exceptions, for example, the *Gentleman's Magazine* (March 1824, p. 260), which assured its readers that those behind the Mexican mining companies were "gentlemen of great respectability and influence," the directors of the joint stock companies formed during the Bubble were all tarred with the same brush. Little distinction was made between those who promoted genuine insurance companies, canals, or railways in Britain, or mining and other companies in Spanish America, and those who sought to defraud gullible investors with promises of vast profits from what were transparently fraudulent schemes. Especially after the Bubble burst with the collapse of share prices from April 1825 onward, and the rapid liquidation of most of the companies thereafter, the newspapers, periodical press, pamphlets, and caricaturists of the day depicted the directors as villains, confidence tricksters, thieves, and charlatans who preyed on the weak and ill-informed, depriving them of their savings and hard-earned pensions. They were "designing knaves," "from dustmen up to peers of the realm inclusive" (*The Times*, 20 April 1826). Those many Members of Parliament who appeared on the company prospectuses were condemned for allowing their names and positions to be used in speculative schemes for which it was widely assumed they were being paid in cash or shares. Cobbett especially pulled no punches. For him, the company promoters were "joint stock blackguards," "monsters of joint stock," "scamps of joint stock companies." The men behind them were "monsters of iniquity," "whose natural propensity is to live upon the labour of others." The Jews and, above all, the Quakers, "that sly, sleek, meek, money-getting tribe," were a frequent target of his venom. Cartoonists were similarly vindictive, especially against the Quaker, Joseph Fry, whose bank—Frys and Chapman—was involved in several of the companies. One cartoon satirizing the spurious "New Combination Pawnbroking Company" was dedicated to the "Quakers, Jews, Heathens, Incendiaries, Usurers and Monopolists, a goodly Fry."

Nor were the shareholders, or proprietors as they were more formally described, spared from this kind of invective. Cobbett, as always, was extreme: "I hold that man to be little short of a downright madman who can expect this parcel of moonshine ever to be turned into gold."[2] Others were less critical. Most shared the views expressed in *The Morning Herald* (11 January 1825) that the shareholders included "many poor, ignorant, deluded persons, whom we may term the unsophisticated part of society and who may be induced to invest their 'little all' in these schemes." The innocent would suffer "while the knaves will keep their horses and carriages." Even more eloquent perhaps was *The Annual Register*:

> Crowds of individuals of every description—the credulous and the suspicious—the crafty and the bold—the raw and the inexperienced—the intelligent and the ignorant—princes, nobles, politicians, placemen, patriots, lawyers, physicians, divines, philosophers, poets, intermingled with women of all ranks and degrees—spinsters, wives, and widows—hastened to venture some portion of their property in schemes of which scarcely anything was known except the name.[3]

But, who were all these people? Were the directors of the Mexican companies all villains as Cobbett and others would have us believe? Were those who put their money into the Mexican ventures as naïve and foolish as the cartoonists claimed? Were they all rich and affluent people who could well afford to take a risk with their capital, or were they from low- and middle-income groups with some small savings, perhaps accumulated over a career in the armed services or the Church? What other social and economic groups were represented? Were there many landowners and aristocrats or were they largely drawn from mercantile, banking, and commercial circles? To what extent were they metropolitan based, reflecting a concentration of wealth, information, and expertise in the City of London? Or did the spread of shareholders extend to the regions and towns across Britain? Were there many small investors, looking for and even dependent on a secure, long-term income? How many were speculators, skilled at playing the market in search of quick profits? In sum, who were the directors and shareholders concerned with the Mexican companies?[4]

It is with these and related questions that we are concerned in this chapter. Identifying the shareholders and their private investments, however, poses a number of problems. In the first place, with one exception, no company registers of shareholders have survived. This may be because no such registers were kept by many of the companies and although from time to time there are references in annual reports to lists held in company offices, these have not survived, or at least have not so far been located. Secondly, and the only registers that have survived illustrate this point clearly, shares were, of course, bought and sold on a regular basis. The names of shareholders changed frequently and the few extant printed lists have many changes entered in them, in ink, and probably by a company secretary. While as we will see later, some shareholders held on through thick and thin, retaining their shares for many years in the hope of eventually retrieving their investments, notably those in the Real del Monte Company, many liquidated their holdings and wrote off their losses. Shareholders, in short, were to a large extent transient and therefore difficult to trace. Another problem is that while the names of shareholders and directors may be found, nothing is known about them. They have left no record of their personal or business lives, social origins, or even address. Hence, although in many cases the extent and value of the holdings of individual shareholders have been ascertained, in others it has not proved possible to find the information.

Fortunately, notwithstanding such problems, it has proved possible to identify a large number of shareholders and almost all the directors of the Mexican companies. Various sources have enabled this to be done. We have already mentioned the existence of some registers. These are from the United Mexican Mining Association. Uniquely among the joint stock companies established to invest in the Spanish American republics, it chose to publish annual lists of its shareholders. The first such list appeared in 1829 under the title: *List of Proprietors of the United Mexican Mining Association, corrected to the 29th July, 1829.*[5] Thereafter, similar, amended lists were published each year until and including 1834, after which as far as is known, no more were produced. Included in these lists were the names of the directors and auditors of the company as well as the names and place of residence of all the shareholders. In addition, the number of votes each shareholder was entitled to cast at general meetings was included. This information enables us to know the approximate number of shares each proprietor held because the votes were determined by the number of shares, as follows:

> Every holder of
> Ten and less than twenty-Five shares: One vote
> Twenty-Five and less than Fifty shares: Two votes
> Fifty and less than One Hundred shares: Three votes
> One Hundred and upwards: Four votes

None of the other companies published their shareholder registers. Similarly, for the most part, the best source of shareholder details, especially in the very early years of the companies, has also not survived. These were the Deeds of Association which, as explained in chapter 2, were drawn up by all the companies at the time they were formed. Shareholders were required to sign the deed, and failure to do so invalidated their shares and the shareholder rights that went with them, such as voting at meetings. Again, as noted earlier, only one original Deed of Association has survived, that of the Mexican Company, formed in February 1825. It is now in the Bancroft Library, University of California.[6] It has the signatures of approximately 220 shareholders with details of the number of shares held by each one.[7] Unfortunately, some of the signatures are illegible, but the remainder provide us with a unique gallery of the shareholders who put their money into a Mexican company in 1825.

Another useful source for identifying shareholders are the annual general meetings and reports regularly published by the different companies. Many of the reports have survived and amidst a wealth of information about the activities, finances, and prospects, there are often details of directors, personnel, and, sometimes, references to shareholders. The main London newspapers usually sent reporters to the annual general meetings, and those shareholders who raised points or asked questions are often named. Similarly, there are letters to the press by irate or other shareholders and while these were usually sent under a pseudonym or anonymously, occasionally, the writer's name is given.

Even more helpful than the sources so far indicated is the existence of share certificates. A substantial number of these, from most of the Mexican companies, have been discovered in a variety of places. Several of the banks that acted for the companies in the 1820s survived the crash and years later were absorbed into the present-day banking conglomerates. Their archives have to some extent been preserved and within them, share certificates belonging to directors or customers of the bank. Lloyds Bank archive, for example, has several of the Real del Monte Company certificates. With such certificates, and government bonds, now much sought after by a growing body of collectors around the world, private sources are revealing other examples that appear from time to time for sale in public auctions. Some collectors have donated their finds to university archives. For example, the University College London archive has a good selection of Bolaños Company certificates. Even more impressive is the collection in the Swindon and Wiltshire Record Office at Chippenham. It has more than 370 of the Real del Monte share certificates, dating from 1829 to 1841 in what is known as the Savernake Estate Collection. These include certificates, all signed, belonging to the original directors of the company as well as many of the other shareholders. How and why this important collection found its way to Chippenham remains an unexplained mystery.[8]

Such "institutional" sources have enabled us to identify, therefore, a significant number of shareholders, but also productive and informative are the personal papers and correspondence of directors and others involved with the companies. A few examples may be given at this point. John Biddulph, of Ledbury, son of a prominent Herefordshire banking and landowning family, was a senior partner in the London bank of Cocks and Biddulph. In February 1826, already a director and shareholder, he became Chairman of the United Mexican Mining Association, a post he held for several years until at least 1836. His meticulous personal diaries kept over many years and now in Herefordshire Record Office include numerous references to the fortunes of the company and his investment in it.[9] His son, John, was a shareholder and another son, Robert, joined the Board around 1840. The Biddulph family remained shareholders until at least the 1860s. Their bank was later absorbed by Martins Bank and subsequently Barclays Bank, and it is in the latter's general archive where we can now find details of the Biddulph family shareholding in the United Mexican Company.[10] It is also noticeable that several other well-known Herefordshire residents, almost certainly friends of the Biddulphs, were shareholders, for example, Thomas Blizard and Elizabeth Bodenham.[11]

Another company chairman whose personal papers, including a diary, have survived was Thomas Fowell Buxton.[12] He was a partner in the London brewery of Truman, Hanbury, and Buxton and entered parliament in 1818 as the member for Weymouth. Increasingly involved by the early 1820s in philanthropic activities, his main reputation came from involvement in the campaign to abolish slavery and as such he was closely connected to William Wilberforce.[13] He was married to Hannah Gurney, whose family owned the Norwich bank of Gurney and Co. He was a founding director and chairman of the Real del Monte and Bolaños Companies, remaining a shareholder until his death in 1845. He was still investing in the 1840s, when he bought twenty-five Bolaños shares, and we know from his diaries that his relatives, for example, his uncle Charles, were also shareholders. While his personal archive is revealing in many respects, however, there is for the most part only passing reference to his business affairs and there is also little information in his memoirs. This is because his son, who edited the memoirs, decided that posterity would not be interested in such mundane matters. He wrote,

> The mining companies alluded to above to which he belonged involved him in considerable loss of property, and their affairs were often a subject of anxiety to him. The details would be of course unsuitable and uninteresting to the general reader.[14]

Another of the Real del Monte and Bolaños shareholders whose personal papers are rather more informative is the zoologist and ornithologist, William Swainson. We have already encountered him in connection with William Bullock and the Mexican exhibitions at the Egyptian Hall. Swainson, never a wealthy man at any time in his life, was an early investor in the mining companies and it is clear from his correspondence that he had great hopes of getting a good return on his money.[15] His wife, Mary, shared his hopes. She bought shares in a Brazilian mining company with a "twelve guinea legacy" on which she received a thirty-shilling dividend. Writing to her father, she was enthusiastic about the future: "Your old head would say realize, but my younger one is filled with golden visions."[16] As for Swainson, he seems to have tried to find out as much information as he could about the prospects of the mining companies before parting with his money. As early as April or May 1824, he sent his friend in Liverpool, T. S. Trail, a sample of Mexican coal, perhaps acquired from Bullock. He asked Trail, a medical doctor but also involved in other sciences, to analyze it and estimate the probability of it being found in large quantities. He apparently was hoping that coal might provide the fuel required for the steam engines that all the mining companies at that time were busily ordering for shipment to Mexico. Trail's results were not positive and he told his friend that large supplies were unlikely to be found.[17] Nevertheless, and notwithstanding the instability of the market where prices in all the mining companies fluctuated considerably, Swainson continued to invest in the 1820s and in the early 1830s, paying calls on his existing shares and buying new ones.[18] He also kept a very close eye on his investment, writing regularly to the Real del Monte and Bolaños company secretaries, demanding progress reports on the mining operations in Mexico and wanting to know why cash in hand was not being distributed as dividends. He asked for and received reports from mining engineers and diligently kept the printed company reports with in some cases detailed and ornate colored maps and drawings of the mines. He certainly received a few dividends from the Bolaños Company in the early 1830s but it seems that before he emigrated to New Zealand in 1841, he disposed of his shares. In a short autobiographical sketch published in 1840, he looked back with evident bitterness on his Mexican mining investments:

> The gradual loss of nearly half my fortune by the utter failure of two of the Mexican mining companies, once the most promising,* would hardly deserve mention, save to warn others against faith in the names and promises of joint stock companies.

To this lament he added the following note:

> that of Real del Monte and Bolaños—ruined by the bad management of the "manager" and his supporters in England. That the proprietors can consent to sink more of their money, under such superintendence, would be incredible—were it not true.[19]

One Real del Monte shareholder, actually recruited by Swainson before his disillusionment, was Peter Mark Roget. He was a doctor by profession—at one time physician to the Spanish embassy—but, of course, is now best known as the compiler of his famous *Thesaurus*. Although he certainly retained his shares until the mid-1830s—there are two of his 1835 certificates in the Trowbridge collection—we do not know what happened to his investment. Initially, he shared Swainson's enthusiasm. In a letter of 7 January 1828, he thanked Swainson for letting him see the information about the Mexican mines, adding that he had taken his advice and bought two Real del Monte shares at £25 premium. Since by then the shares were fully paid up at £400, Roget presumably paid the substantial sum of £425 per share or £850, plus his broker's commission. Fortunately, he went on to say, the premium had since increased to £50 to £60 per share. He asked Swainson to continue to send him any more information he had about the mines. After all, he said, "I am now an Adventurer." Two years later, in March 1830, he bought more Real del Monte shares and again writing to Swainson to ask to borrow a map of Mexico, he noted, "I see they maintain their price at a small premium which augurs well."[20]

The Strutt family of Derbyshire were also within Swainson's circle of friends and acquaintances. The Strutts were wealthy mill owners and textile manufacturers whose family fortunes had been founded by Jedediah Strutt (1726–1797). In December 1827, his grandson, Jedediah Anthony Strutt, bought ten shares in the United Mexican Company, at a discount of £18 per share on the paid-up price of £35, thus paying £17 per share for a total of £170 plus commission and stamp duty at £4 10s. A couple of weeks later, he returned to the market and bought five shares in the Mexican Company. These were also at a heavy discount. The paid-up price on each £50 share was £23 but Strutt bought at a discount of £14 10s, thus paying no more than £8 10s. per share plus duties and commission—in sum, his Mexican Company shares cost him £42 10s. His initial investment, therefore, at a total of £217 was not a substantial sum for a man of his means. Unlike Swainson, he seems never to have lost faith in the companies, holding on to his shares for the rest of his life. Over the next twenty-odd years he faithfully paid every call or financial demand made on him by the Mexican Company—it issued calls for cash from shareholders more or less every year—and in his papers there are receipts for approximately £186 paid in regular installments between 1828 and 1851. His son and heir, George Strutt, was unaware of the investments and it was only when the United Mexican Association wrote to its shareholders in 1856 to request further payments that his father's holdings came to light. Detailed investigation by the executors of his will found that none of the six dividends paid between 1845 and 1853 had been claimed. Another member of the family, Joseph Strutt, was also a shareholder in the United Mexican Association, but the details of his financial dealings have not been seen.[21]

Finally, mention may be made of another very well-known name who chose to invest in the Mexican companies. This was the landowner and photography pioneer, Henry Fox Talbot, of Lacock Abbey, Wiltshire. At first, he seems to have been reluctant to risk his money. In a letter of 17 January 1825 when the Bubble in the stock market was rising to its peak, his advice to his half sister, Horatia, who seems to have asked about buying shares, was clear: "Anybody might have made their fortune this year in the Mexican mines, the time is gone by now, and those who try the concern now will very likely be ruined."[22] By the end of that year, 1825, however, he seems to have changed his mind and bought one share in the Anglo-Mexican Mining Company. Shortly afterward, in February 1826, he bought two more, paying £25 per share, and was immediately called upon to remit another £5 per share to pay a call issued by the Company. This demand does not seem to have bothered him unduly because in October 1826, he added to his holding in the same company with the purchase of five more shares. Eventually, he held ten shares. He kept them until February 1829, when he instructed his broker to sell, and they fetched £20 each or £200 in total, less fees and commissions of £1 10s. Although we do not know exactly what he paid for the shares, it seems certain that he lost money on them.[23]

These sources—shareholder registers, a Deed of Association, published company reports, newspaper accounts of meetings, bank and other archives, extant share certificates in private and public collections, and the personal records, and especially correspondence of individual shareholders—have provided the identities of approximately 1,200 of the shareholders in the various Mexican companies. For about half of these, information on addresses and occupations has been found in contemporary post office and commercial directories, together with a variety of other sources. These details reveal that the largest groups of shareholders were from the mercantile and financial sectors with merchants (148), stockbrokers (65), and bankers (40).[24] Also prominent were military officers (23), with ranks ranging from sergeants to generals and vice admirals. There is also a noticeable number of Members of Parliament (26) and titled aristocrats (29). Women (56), both married and spinsters, held shares in all the companies. Smaller groups included brewers (16); clergy (12); medical practitioners (8); sugar refiners (5); insurance brokers (4); and jewelers (4); solicitors/lawyers (5); chemists (3). Finally, in what we may term miscellaneous trades/professions, a myriad of activities is represented by booksellers, wholesale stationers, coal factors, grocers, printers, wallpaper manufacturers, watchmakers, wholesale cheesemongers, bill brokers, hosiers, surveyors, and architects.

Looking first at the merchant group, it is evident that many prominent individuals and partnerships in London's mercantile community are represented. To give just a few examples, there is John Diston Powles, Charles Herring, Burmester Brothers, Burnand, Lewis, and Co., William and James

Burnie, Cohen Brothers, Collinson, Stanley, and Co., Dartez Brothers, Edwards and Co., Exter Brothers, Green and Hartley, Hanson and Co., Henry Alexander and Co., G. W. and S. Hibbert, Thomas Horsley, Hullett Brothers, and George Rougement and Co. Even better known perhaps are the four Cazenoves (Henry, James Jr., John, and Philip), shareholders in the Catorce Company and the United Mexican Company. At this time they were in the process of starting the merchant bank that still exists to this day. George, John Aldred, and Richard Twining, all of the now famous tea merchants and also banking family, had shares in the Mexican Company and the United Mexican Company. George Sandeman, founder of the wine and port wine company and a partner in Sandeman, Gooden and Forster, was an investor in the United Mexican Company. Another shareholder in the same company was Money Wigram, merchant, ship-owner, and director of the Bank of England. His father, Sir Robert Wigram, is famed for his commercial success as well as having fathered twenty-three children, nineteen of whom survived into adulthood and enjoyed considerable success in their careers. John Lavicount Anderdon, described in the *DNB* as a writer on angling and devotional subjects, was a partner in the West Indies merchants, Manning and Anderdon, and his shares were in the United Mexican Company. Finally, Henry Hunt, radical pamphleteer and journalist, known in his day as "Orator Hunt" but also a successful merchant, owned shares in the Real del Monte Company.

These few individual examples, and dozens more could be given if space permitted, confirm that the Mexican companies attracted support from many already successful and experienced merchants who were certainly not to be duped by false promises and an extravagant prospectus. Faced with a broad choice of places in Britain, Europe, and the Americas in which to invest, they chose Mexico and they confidently expected to get a good return on their money. With their ready access to the Stock Exchange and to brokers, it is no surprise that London-based merchants predominate but there is no shortage of their counterparts and other businessmen from many other parts of Britain. Again, to give just a few examples, there was Thomas Were Fox, a serge manufacturer of Plymouth; Samuel Guppy, a Bristol sugar refiner; Edward Bevan, a Bristol merchant; Joseph Read, a wine merchant at Cheltenham; Robert Tindall, from a prominent shipbuilding family of Scarborough; John Wilkes Unett, Birmingham solicitor, later credited with the development of the seaside resort of Filey; John Wylie, a Manchester merchant. Many other names could be added to the list and the same applies if we look at some of the other groups. There were, for example, twelve clergymen, such as the Rev. Thomas Coles, Bourton on the Water; Rev. Thomas Oliver Rogers, Weston-super-Mare; Rev. Sir William Murray, Lofthouse, Yorkshire; Rev. James Capper, Wilmington, Sussex. These were all United Mexican Company shareholders but the other companies also attracted the clergy. One surviving share certificate from 1824 belonged to the Rev. John Faithful Grover Fortescue, of Rowell Vicarage, Essex. He bought a single £100 share in the Anglo-Mexican Mining Company on 20 December 1824, for which he paid £10 cash as the first installment.

Most of the major London stock-broking firms were shareholders, although it is often not clear whether they were acting for clients or on their own behalf. Bankers and brewers among the shareholders were more dispersed around the country. London-based bankers included influential figures in the financial community, such as Sir John Lubbock, Frederick Huth, several members of the Smith family of Smith, Payne and Smith, William Fry of Frys and Chapman; three Bosanquets of Bosanquet, Pitt, Anderdon and Co. Provincial bankers were represented by Henry Alexander (Ipswich); Timothy Cobb (Banbury); James Heygate (Leicester); and Christopher Rawson (Halifax). Norfolk and Norwich were strongly represented by Henry Birkbeck; Thomas Cresswell; Anthony and Thomas Hudson and their partner, Sir John Harvey; as well as Samuel Hoare and Samuel Gurney, both of well-known banking dynasties. Norfolk brewers were also major investors. There was Michael Bland, Thomas Fowell Buxton, and several members of the Martineau family. Robert Cobbold was an Ipswich brewer with shares in the United Mexican Company and Thomas Meux of the London Meux, Reid Brewery was a Mexican Company shareholder.

Turning to the twenty-six Members of Parliament who were directors of the companies or held shares in them, no particular political persuasion is evident. As the caricaturists often pointed out, they came from "Whigs, Tories, Radicals, and Saints." We have noted earlier, however, that much of the contemporary criticism of the joint stock companies was directed at Members of Parliament whose names appeared on the prospectuses. They were accused of being opportunists, taking advantage of their status to defraud the public, and to profit by share speculation in the free shares they were given in return for their cooperation. There seems little doubt that there were instances of such guinea pig or decoy directors in some of the companies, and when the crash came, allegations of corruption and malfeasance were soon heard.[25] Examination of the personal profiles and careers of those parliamentarians involved in the Mexican companies does not provide any evidence to justify the allegations. All appear to have been serious investors and genuine participants. Most were already wealthy men with a mercantile or financial background, and in several cases they were still pursuing their private business interests. Stewart Marjoribanks, Matthias Attwood, and Thomas Wilson all had a mercantile or business background. William Ward, a director of the Bank of England, was a financier and expert in foreign exchange markets. He was a cousin of Henry George Ward, Britain's chargé d'affaires in Mexico from 1825 to 1827. Active in several joint stock companies in 1825, he is probably best remembered for his support of the game of cricket. An enthusiastic and talented player himself, he purchased the lease of what is now Lord's cricket ground in the same year for £5,000 from Thomas Lord, who had planned to allow building developments on the land. Among the MPs who were shareholders but not directors, there was John Wells from a wealthy family of shipbuilders; Walter Boyd, a financier with extensive European interests; and Maurice Fitzgerald, from an Irish landowning family; Charles Poulett Thomson, son of a London merchant engaged in trade with Russia and the Baltic; and Henry Baring, partner in the family bank. Other examples confirm the picture that emerges of men with personal wealth and wide experience in mercantile and financial affairs. It appears improbable that they would have allowed their names to be used in projects that were known to be intentionally fraudulent or have invested their money in ventures with no prospects of success.

It is also noticeable that among the twenty-nine shareholders with aristocratic titles, only five were directors. They were Sir Stephen Sharp, a director of the Mexican Company who held 210 shares in it. He died in 1826 and hence had little time to participate in its affairs. Sir Gore Ousely was a director of the General Pearl Company. He was a professional diplomat and noted oriental scholar and may be one example of a guinea pig name since as far as is known he had no relevant expertise to offer the company. In the same category were possibly three senior naval officers—Vice Admiral Sir Henry Blackwood; Rear Admiral Sir William Fahie; and Vice Admiral Sir Malcolm Pulteney—whose names appear on the boards of steam navigation companies. On the other hand, Rear Admiral Sir Charles Ogle was not a director but he did have shares in the United Mexican and Real del Monte Companies. Among the other titled shareholders, there were three women; the Dowager Viscountess Anson; Hon. Miss Frances Catherine MacKenzie; Lady Beauchamp Procter. There are fifteen titled names in the United Mexican Company share register, and another six were Real del Monte shareholders. All appear to have been normal investors rather than lending or selling their name for immediate profit. They include Viscount Eastnor; Earl of Falmouth; Counts Henry de la Pasture and James Alexander Pourtales Gorgier; assorted baronets such as General Sir James Affleck; Sir Henry Chamberlain who gave his address as "Foreign Office, Downing Street"; Sir James Hanham, and Sir John Stuart Hippesley.

Another category is the women shareholders. Fifty-six have been identified of whom forty-two are listed as spinsters (Miss) and eleven as married (Mrs.), with three titled aristocrats. There are several of what appear to be family groups, for example, Misses Francis Mary, Elizabeth, and Jane Elizabeth Blackwell who all lived at 6 Seymour Place, Mayfair. They each had shares in the United Mexican Company and Jane Elizabeth also in the Real del Monte Company. The Hodge family, Sydenham, Kent, was similar: Edward Cooper Hodge and Misses Anna Maria, Caroline, and Eleanor Hodge, all United Mexican Company shareholders. Also of Sydenham, there was Mrs. Mary Mayow and Miss Frances Wynell Mayow, presumably her daughter. In all there are ten family groups in which more than one female member had bought shares in the Mexican companies.

Among the miscellaneous group, there was Rudolph Ackermann. In addition to investments in various Latin American bond issues and South American mining companies, he spent £80 on shares in the Tlalpujahua Company and may also have put £200 into Bullock's ill-fated mining venture. He had met the Tlalpujahua Company's promoter, Rivafinoli, and he was apparently impressed with the promising future he depicted for the company. Furthermore, his son, George, who traveled to Mexico in March 1825 to open up a branch of the family bookselling and publishing business, had only recently completed three years of study of Mining and Mineralogy at the University of Freiburg in Southern Saxony. He sailed to Mexico on the same ship as Waldeck and other Tlalpujahua Company employees.[26] There were also four Oxbridge dons, namely, Samuel Fennel and Joshua King, Queen's College, Cambridge; William Heathcote Unwin, Trinity Hall, Cambridge; Edward Field, Queen's College, Oxford—all had shares in the United Mexican Company. Other notable names include Charles McGarel, fourth son of an Irish innkeeper, sugar merchant, and multimillionaire, who had shares in the Mexican Company. In later years, he was to be the largest individual holder of Mexican bonds and chairman for a time of the Mexican Bondholders' Committee.[27] In contrast, in terms of wealth and influence, there was Isaac Whitwell, a wholesale grocer and tea merchant in Kendal, a small provincial town that provided a significant number of investors who saw opportunities in the new Spanish American republics.[28]

Finally, in this group survey of the shareholders, mention must be made of foreign investors. Twenty-two are listed as living overseas; twelve in Geneva; two in Amsterdam; two in Paris; and others in Dublin, various places in Germany, Modena (Italy), and Guernsey. Again, most of these investors held shares in the United Mexican Company and at least in one instance, three members of the same family, namely, Alexis, Antoine, and William Saladin, all of Geneva.

Turning to the geographical distribution of the shareholders, it is evident that the majority lived in London or the metropolitan area where it was certainly easier and quicker to become involved in the stock market. On the other hand, it is also the case that many provincial towns and villages in all parts of Britain from Brighton in the south to Staindrop, near Darlington, in the north, had shareholders in one or more of the companies. To list them all is unnecessary and what follows is only an alphabetical sample: Bath, Bristol, Cheltenham, Halifax, Lancaster, Leeds, Lincoln, Lothbury, Manchester, Plymouth, Sheffield, Swindon, Warwick, and Worcester. There were a few in Scotland, mostly Edinburgh, and also some in Dublin. Some places provided an unusual concentration. Liverpool, for example, was well represented with at least twelve, including Thomas Headlam, an insurance broker; Joseph Hornby, a merchant; Christopher Jones, a jeweler and silversmith; Charles MacMurdo, a glass merchant; and William Lassell. He was a very successful brewer who is now remembered in the *DNB* for his contribution to astronomy, an activity on which he apparently lavished many thousands of pounds. His shares, between ten and twenty-five, were in the United Mexican Company.

Another region that provided more than the usual number of investors was Norfolk. We have seen how Roger Allday Kerrison and his family retained control of the El Bote Mining Company, deriving handsome dividends from it for almost a century. It remained throughout a private business largely confined to the Kerrison and Placci families, but there were at least twenty other Norfolk families who invested in the Mexican joint stock companies. These included landowners, brewers, and bankers, as well as one woman, Miss Jane Maria Scott, of Norwich. The prominence of Norfolk was particularly related to the Real del Monte Company. As we noted earlier, it is the best known of all the joint stock companies formed to invest in Mexico. But notwithstanding all the research done into its history, it is surprising that no attention appears to have been paid to its London management, that is, the directors, nor to the shareholders who provided the millions of pounds that were spent on the mining operations in Mexico. Information on the identity of the shareholders has not been easily available but the discovery of almost 400 share certificates, especially the Chippenham collection, now enables us to know who provided much of the money. Before turning to the individual shareholders, however, there are certain features of the company that need to be explained. It originated in Norfolk and almost certainly was the inspiration of a group of Norwich-based families, friends, and business associates. John Taylor, the mining engineer and the company's general manager for many years, was clearly the main promoter. He was from a prominent Norwich family who were closely associated with the Martineau family. They also had extensive brewing and banking interests in Norwich and they had investments in Cornish mines. John and Joseph Martineau were among the founding directors of the Real del Monte and Bolaños companies. Peter Martineau was one of the auditors and Martineau and Malton, another branch of the family, were appointed the company solicitors. Another friend of the family and also from a Norfolk banking and brewing dynasty was Michael Bland. His share certificates survive in Norfolk Record Office.[29] They reveal that he was one of the earliest subscribers, buying on 30 November 1824 five shares numbered 47–51 on which £70 each was paid. He made further purchases in following years and paid the calls as required. Samuel Hoare, Quaker banker, related to the Bland family, was another Norwich investor as was Thomas Jex Blake, one of Norfolk's principal landowners. Finally, we have already explained that the chairman of the company for many years was Thomas Fowell Buxton, also from Norwich. His brother-in-law, Samuel Gurney, a member of Norwich's premier banking family and also a Quaker, joined the Board of Directors soon after the company was founded.

The Norwich connection was clearly very strong and it is noticeable that most of these families were Quakers and several were related through marriage. It is no surprise, therefore, that another of the founding directors was William Fry, a partner in the Quaker Bank of Frys and Chapman which became the company's bankers. The Quaker presence was clearly important for the reputation of the company and may well have contributed to the initial high premium that the shares reached. The Quakers, although the target of Cobbett and critics of the joint stock companies, still enjoyed a good reputation for their financial acumen and honesty:

> They supported each other in public and private life, and as they eschewed extravagant living and regarded hard work as a Christian virtue, they became reliable partners, and in particular, reliable bankers.[30]

The remaining directors are not known to have Norfolk or Quaker connections but they were all distinguished men in their own fields. Francis Baily was a stockbroker and a Stock Exchange manager, now remembered mainly for his later work as an astronomer; Thomas Brown was a merchant; Major John Gurwood was an army officer with a distinguished career in the peninsular war and at Waterloo; Thomas Colby was a surveyor and army officer in the Royal Engineers. Edward Hawke Locker was an experienced administrator, author, accomplished watercolor painter, and in 1824 was appointed civil commissioner of the Royal Naval Hospital at Greenwich. Finally, there was S. F. Wilde and J. H. Shears, about whom nothing is known. The status and importance of these directors is indicated by the fact that five of them—Buxton, Baily, Colby, Locker and Gurwood—are all deemed worthy of inclusion in the *DNB*.

These Norfolk bankers and brewers resolved early in 1824 to set up a private company with just 500 shares and we know from the court case with Kinder over the distribution of the shares that Taylor had control of 300 and Kinder, of 200. The shares were distributed to the founding members and their friends in Norwich. They were not intended to be placed in the market and according to John Taylor, very few were ever offered for sale. In a long letter which he wrote on 4 September 1826 for publication in the newspapers but which never appeared, he explained that "the shares were limited to a small number, that speculation in them might be discouraged, and they were fairly divided by the parties themselves before a Director was chosen."[31] Indeed, as far as he knew, the notorious price of £1,300 to which the shares were said to have once risen, had only existed in one or two instances. He went on to emphasize that the original promoters of the company had not sold their shares—"they have been purchasers, not sellers"—and that they had nothing to do with the speculation in the few shares that had come into the market. The extant share certificates confirm Taylor's claims. The first or original shareholders such as Buxton, the Martineaus, Bland, and others held on to their shares for years, paying the regular calls and subscribing to the new issues throughout the 1820s, 1830s, and early 1840s.[32]

Who, then, were the shareholders in the Real del Monte Company in addition, of course, to the founding directors? Divided into groups, the picture is much the same as for the generality of shareholders in all the Mexican companies: bankers (14); merchants (12); brewers (8); stockbrokers (7); women (5); titled nobility (4); clergy/doctors (3); military officers (3). Members of Parliament were well represented with at least seven shareholders and there were assorted other professionals such as the zoologist, Swainson; the editor and printer, Abraham John Valpy; and Horace Twiss, lawyer, "wit and politician," according to his *DNB* entry. Employees of the company were also investors. James Vetch, one of the first commissioners or managers sent out to Mexico, put all his savings into the company and was firmly convinced that he had made a sound investment. He told John Taylor that "I see nothing but the most clear prospects of success."[33] Thomas Rule, almost certainly related to another commissioner, John Rule, was also an investor. There were many other eminent names among the shareholders in addition to those we have already mentioned. Sir John Easthope, for example, was a successful stockbroker and Member of Parliament for St. Albans (1826–1830) and other constituencies in later years. In addition to his shares in the Real del Monte Company, he was also a shareholder and later chairman of the United Mexican Company. Other notables include Sir John William Lubbock, leading London banker; Sir Robert Price, Welsh iron manufacturer, Member of Parliament, shareholder, director, and later chairman of the company; James Foster, Quaker and partner in the stockbrokers, Foster and Braithwaite; Rear Admiral Sir Charles Ogle, wealthy and successful naval officer; Charles Stuart, Baron de Rothesay, senior diplomat involved in many important missions; Sir George Elliot, naval officer, politician, and later Member of Parliament for Roxburghshire.[34]

In general, the Real del Monte Company shareholders, with a few notable exceptions like Swainson, were clearly from the more affluent mercantile and commercial sectors who could afford to buy the first tranche of shares which cost the considerable sum of £400 when fully paid up. This fully paid-up price was reached by the end of 1825 and while perhaps a few had sold to benefit from the high premiums the shares fetched earlier that year, many held on to their shares and paid the regular calls to the full price of £400. The original shareholders invested, therefore, substantial sums of money in cash in the company. The first annual accounts, dated 12 February 1825, show that £58,800 had been raised and spent in the first year. The first two calls on the shares of £20 and £50 respectively were paid by every shareholder and only twenty-four failed to pay the third call, also of £50.[35] In later years shares were offered in fractions—one-quarter, one-eighth, one-sixteenth—in an attempt to reduce the amount of cash each shareholder was called upon to find.

Among the most loyal shareholders was the company chairman, Thomas Fowell Buxton. Unlike so many of his Norwich friends and colleagues, he was not a Quaker, even though his mother was an active member of the sect. Buxton was baptized in the Church of England, thus adopting his father's Church, but he always retained a strong attachment to the Quakers and regularly attended their meetings. He also regularly visited his sister Anna. She had married a Quaker, William Forster, and they lived at Bradpole, near Bridport, where there was a strong Quaker community.[36] Members of the Gurney family were also frequent visitors.[37] Whether by coincidence, or by the influence of such visitors, and especially Buxton, who was described in parliament as "the great champion of all sorts of joint stock companies," the county of Dorset, and in particularly, the town of Bridport had the highest concentration of shareholders in all the Mexican companies.[38] They provide a useful microcosm of the socioeconomic groups who were tempted into the market by the appearance of both the Latin American bonds and the diverse mining and other companies. At least twenty-seven Dorset residents held shares in Mexican companies. They include Dr. John Carnegie of Wimborne, who had between twenty-five and fifty shares in the United Mexican Company and retired army officer, Colonel John le Mesurier, Alderney Cottage, near Poole.[39] William and George Davey were Dorset farmers with shares in the same company. George Filliter was from a well-known family of solicitors in Wareham and Sir James Hanham of Wimborne was a large landowner in the area. Women were also represented by Miss Jemima Lee; and Misses Augusta and Mary Lloyd, all of Lyme Regis. Among others were Henry Davis, Swyre; William Davis, Netherbury; John Kingston Galpine, Blandford; John Goss and Thomas Salter, Poole; and Henry Hill, Wimborne. All had varying numbers of shares in the United Mexican Company.

The county of Dorset, therefore, was strongly represented in the shareholder registers and nowhere more so than the small coastal town of Bridport. With a population of perhaps 2,000, including a strong presence of Quakers and other non-conformist groups, Bridport's main economic activities involved wool and rope making. It had at least twelve shareholders in the Mexican companies, most of whom were either related or had business or social links. Again, all had shares in the United Mexican Company. William Battiscombe had between fifty and one hundred shares; Miss Sarah Bools, from a prominent shipbuilding family, had over one hundred shares and she also had substantial investments in Chilean bonds; Joshua Carter was from a leading Presbyterian family; Miss Mary Coppock, possibly from a family of twine merchants, had twenty-five shares; Misses Maria and Mary Fathers, one hundred shares between them; Misses Sarah and Ann Golding, daughters of a Bridport merchant, owned shares but the number is not recorded; John and Joseph Hounsell, rope and twine manufacturers and like Joshua Carter, members of the Presbyterian Church, between twenty and fifty shares.

The central figure in the Bridport shareholders' circle was Thomas Collins Colfox. He was a successful and wealthy woolstapler and general merchant with business connections throughout Great Britain. Even though Bridport was a small town isolated to some extent from the center of political and commercial affairs—it cost 10d (£10 in our terms) to send a letter to London—Colfox kept himself informed of both national and international events by reading the London dailies, especially *The Times* and *The Morning Chronicle*, to which he often referred in letters to family and friends. Like many of his contemporaries, he despised Spain and its monarch, Ferdinand VII, and fervently supported the liberation of the Spanish American colonies from what he saw as colonial despotism. Referring to a report of the fall of Lima he had read in *The Morning Chronicle*, he wrote, "I hope in God it will prove true, and that the whole of America will be free and independent."[40]

He was also a frequent traveler to London, where he learned of the latest news of developments in international affairs and in the stock market. With ample surplus cash resources from his business activities, he was one of the early speculators in the foreign bond market, investing in Spanish bonds issued by the liberal governments of the early 1820s and in both Colombian and Chilean stocks. By 1822, he had £23,000 in "Chileans," and the following year, he instructed his broker "to send me down" another £8,000 of "Mendez bonds."[41] His broker was Robert Hills and he kept him informed by letter of movements in the market and political developments, not always accurately. In a letter of 7 February 1822, for example, Hills wrote that Spain was about to recognize the independence of her former colonies and that commissioners were to be sent to each new nation and treaties of commerce signed. Colombian bonds, he added, were at 67 and about to go up to 75.[42]

We do not know to what extent Colfox made a profit from his speculations but he did counsel his son in October 1822 that "speculators need strong nerves."[43] He also advised his family and friends on the right time to buy and sell. In a letter to his daughter, Harriet, written from London in January 1824, he told her to tell all their friends in Bridport to sell their Spanish stocks which were going down but to buy Chilean which were going up. In the same letter, he revealed his interest in mining shares:

> The Mexican mine shares are undergoing some change—there is to be an incorporate company formed with Directors and is expected to be a good thing. If anyone wants to buy, tell them the Company is to be formed and run

> from London, with Mr. Alaman, Finance Minister in Mexico and, it seems, a very respectable man. This is the plan but not yet settled, I believe. I will have an option on some shares. I spent a long time with Messrs Hullet, truly respectable men, and discussed the prospects with them.[44]

The company he was referring to was the United Mexican Company. He bought between twenty-five and fifty shares in it and both his daughter, Harriet, and his son, William, took his advice and did the same. Between them, they had in excess of 150 shares that they probably obtained at the issue price of £5 per share or £750 in total. Additional calls were soon made on them—£5 in October 1824 and £10 in January, 1825—with later demands taking the price to the full level of £40. Their outlay was in the order of £6,000. Even their "faithful servant, Betty Sprake" bought some of the shares although we do not know how many.[45]

The Colfox family and their servant, Betty Sprake, retained their shares until at least 1834 but there is some indication that Thomas had intended to sell when the market was rising, which was between November 1824 and March 1825. We do not know his reasons but he failed to sell and the value of the shares suffered a rapid fall in the final months of 1825. Again in London in 1826, he wrote to Harriet and lamented the missed opportunity: "Oh! that we had all been so fortunate as to have got rid of all our South American securities a year ago." It would have made, he continued, an astonishing difference to their property and comfort but "when one looks back, one beholds compleat [sic.] infatuation to have laid hold of many." Even so, he added, the broker Hills remained convinced that the "United" would one day be "a good, a very good, thing" and he had been persuaded by him to buy even more shares in the company but "the folly of this is now evident."[46]

In the same letter, he referred to Miss Bools, family friend and also a shareholder in the United Mexican Company, perhaps as a result of Colfox's own early enthusiasm. She had put much of her capital into the market and had been "exceedingly wrong" to do so, as Colfox put it. Now she was faced with the need to find extra cash to pay the calls on her United Mexican shares and she did not have the money. Her mistake, he said, had been not to sell "in due time." Now he advised her to sell her Chilean bonds and use the proceeds to pay the calls. In the meantime, he promised to ask Hills to find out if the directors of the company would allow her a month or two to raise the money. We do not know if Miss Bools took the advice but she certainly managed to retain the shares, her name appearing on the United Mexican Company register until at least 1834.

As for Thomas Colfox, his correspondence reveals that he continued to try to keep abreast of events and changes in the market. Writing to his broker, Hills, in September 1826, he asked if there were "any cheering prospects" regarding the United Mexican Company. We do not have Hills's reply but it would certainly have been encouraging. In 1827, he published a pamphlet in which he painted a glowing picture of the prospects both of Mexico as a whole and of the United Mexican Company. The country, he insisted, had such unlimited and unexploited natural resources that its future prosperity was guaranteed. The loans and investments that British people and companies had made were not lost but rather "they are seeds sown in a fruitful soil." As for the United Mexican Company, it had excellent mines and three other important qualities: "funds, good management, and scientific and practical skill." The shareholders could expect rich returns on their investments.[47]

Also in 1827, accompanied by his son, Colfox went to London to find out what he could about his Colombian bonds and whether there was any prospect of dividends being paid. A few weeks later, back in Bridport, he wrote to another friend in London to ask about Mexican bonds and if there was any information on "Mexican concerns generally." Finally, while he was not impoverished by his investment in Latin American bonds and mining company shares, his wealth was certainly diminished. Replying to a request for money in March 1832, he regretted that he could only contribute £25 because "my property is so locked up in foreign securities," which were not paying interest or dividends.[48]

The Dorset and Bridport shareholders provide us with an excellent illustration of the sort of people who were persuaded to put their money into the Latin American and especially the Mexican companies. While in the metropolis wealthy bankers, stockbrokers, merchants, and brewers were evident in all the companies both as directors and as shareholders, in the towns and cities across Britain, the investors came largely from the middle class. They were the local doctors, solicitors, farmers, merchants, shopkeepers, retired civil servants and military officers, clergymen, and widows and their unmarried daughters. In general, they were people who were certainly not poor but who perhaps lived off a relatively small amount of invested capital. With British interest rates, especially on government stocks like the traditionally safe haven of the consols, in decline, they looked elsewhere for an improved return on their capital. Attracted by the lure of the legendary Mexican silver, and by the glowing prospects they read in the company prospectuses, they thought, as Colfox put it, that they were onto "a good thing." Although it is difficult to cite the precise nature of the influence, there seems no doubt that the pro-Mexico propaganda, ranging from Bullock's exhibition and book to all the other sources of that Mexicomania which we have described in an earlier chapter, spread across Britain to convince hundreds, if not thousands, of people to enter the market. But while there may well have been some inexperienced and naïve investors, as the joint stock critics of the time insisted, many of those we have identified were informed and cautious with their money. Like Swainson and so many others, they expected that dividends would follow when British expertise, technical prowess, and steam power were brought to bear on the hitherto unexploited or neglected mines across America but above all, in Mexico. But they knew, and were often told as much by the company directors, that instant or quick profits were unlikely even though the dream of bonanzas was always present. Instead, they were willing to be patient and wait for what was to be a very substantial investment of their money to produce the promised dividends. Hence, they held on to their shares for years. In all the companies that survived past 1830, at least some of the original shareholders continued to pay the calls and provide the additional cash demanded by their mine managers in Mexico. As for the directors, it seems quite clear that in the case of the Mexican companies, they were far from the image, portrayed by some newspapers, caricaturists, and other commentators of the time, of scoundrels and swindlers intent on defrauding the naïve public of their savings. With only a few exceptions that we shall reveal below, all the directors could be said to have been respectable businessmen who invested much of their own money in their companies and continued to do so for many more years after the market Bubble of 1825 had burst. Looking back in 1849 at the events of the 1820s, John Lee Comstock condemned "the ignorance and folly on the part of a considerable portion of the public" but he added, "It appears that the original promoters of these mining companies were misled by their ignorance rather than a desire to deceive the public."[49]

Most of the criticism leveled at the promoters of the joint stock companies centered on the charge of speculation. It was said that the directors and others behind the companies made large profits by manipulating the market and their company shares to their own personal advantage and at the expense of the innocent investor. The basic method employed was explained by John Cam Hobhouse, Member for Westminster, in a parliamentary debate held on 16 March 1825 on the subject of the Pasco-Peruvian Mining Company. Let us suppose, he said, that a new company with 10,000 shares was planned. The original promoters—he used 50 as an example—each took 200 shares and paid a £5 deposit per share or £1,000 in total. A prospectus was then published and prospective buyers were warned that the shares were already at a premium and cost £16. An original promoter could then sell in the market 66 of his shares for £1,056. He now had a profit of £56 on his original £1,000 investment and he still retained a further 134 shares "without the risk of a farthing." If he sold at £16 he would pocket £2,144 but if the shares rose, as some had done to £500 or £700, the profit would be as much as £80,000 or £90,000. Vast sums of money were thus being made, he insisted, by unscrupulous promoters, at the expense of innocent investors who believed the prospectus propaganda and the promises of allegedly respectable businessmen and politicians whose names appeared on the company publicity.

Demonstrating now that such "puffing up" of the market, to use the contemporary phrase, actually took place is, of course, difficult. While it almost certainly did occur with some of the obviously fraudulent new companies that were announced, no evidence has been found that such tactics were used to promote the Mexican companies. On the other hand, there was certainly some speculation in their shares, and it is those speculators who form the final group in our survey of the shareholders. There are both individual and institutional examples. Among the individuals who speculated and lost was Dr. Mackie. We know that he was involved in several of the mining companies. He was said, in a private letter of 6 July 1825, to have become "so rich by the rise of the South American securities of all kinds and particularly of the mine shares that he does not wish to trouble himself further about money matters."[50] Apparently he continued to speculate, however, and two years later, Dr. Mackie found himself in the King's Bench Prison for debt. The artist, Benjamin Haydon, met him there and noted in his diary, "it was interesting and curious to talk to Dr. Mackay [sic.] who had lost £40,000 in the Stock Exchange (which he had amassed in Mexico, by a long life of labor) in mining speculations."[51] Another less than successful speculator in mining shares was Benjamin Disraeli. His stockbroker's records and his own correspondence in 1824 and 1825 reveal his involvement in several of the companies, buying and selling shares on a regular basis for himself and his partners, including the publisher John Murray, to the extent that at one point, as he put it, "I found myself in possession of shares in all the great mining Companies to the amount of many thousand pounds." He claimed to have read every book written on mining in America, to have had conversations with secret agents of the companies, and to have privileged access to information from Mexico such as "secret reports which have not even been seen by many of the Directors themselves."[52] Even these apparent advantages, however, and his own propaganda pamphlets on mining, written for the benefit of his mentor, John Diston Powles, as well as himself, were to no avail. He consistently bought and sold at the wrong time and by the end of June 1825, he had managed to lose about £7,000.[53] Finally, there was the case of Sir William Adams/Rawson, oculist to the prince regent and enthusiastic advocate of investment in the Mexican mining companies. All sorts of rumors were attached to his involvement in the stock market bubble. He was said to have refused an offer of £60,000 for his shares in the Real del Monte and other Mexican companies, because he expected them to rise even further. Then, the bubble burst, prices collapsed, and he lost everything. He committed suicide in his own house.[54]

Rather more successful speculators included the Grenfell family. Pascoe Grenfell, Member of Parliament for Penryn in Cornwall, owned a large tin and copper smelting business, and he was a director of the Brazilian Company and the Colombian Mining Association. His sons, Pascoe St. Leger and Charles, were both United Mexican Company shareholders. Charles also had substantial holdings in the Anglo-Mexican Company, noting in his private ledger that "10 shares cost me £1,824 10. 0." He also bought and sold shares in several of the other Latin American companies, notably the Brazilian Company, the Colombian Mining Association, and the Colombian Agricultural Association, in which for a time he held some 250 shares.[55]

Another shareholder who certainly seems to have made good profits was General Wavell. His private papers contain a number of documents relating to his shareholding activities and although their precise meaning is by no means always clear, it appears that substantial sums were made from dealing by him or on his behalf. For example, his father told him in a letter of 12 July 1825 that he had been allocated 750 shares in the United Mexican Company, presumably in return for transferring rights to mines he controlled in Mexico. His broker was Robert Mills and he sold 150 of the shares on 9 August 1825 for £23 each, or a total of £3,450. Further sales in the market yielded over £12,000 and it is interesting to note that among the buyers and clearly speculators in the shares were John Easthope, Thomas Kinder, Sir Robert Wilson, William Ward, and Jacob Ricardo. According to Wavell's father, 750 shares each were also transferred to Sir William Adams and the Exter brothers (John and Richard), again presumably for services rendered.[56] Lucas Alamán was also the beneficiary of 750 shares in the United Mexican Company in return for his help in acquiring mining concessions. Although there is a power of attorney in his papers which he was supposed to sign and return to London to facilitate the sale of the shares in the market, it is not clear if and when they were sold.[57] Rumors of other speculators and their successful or unsuccessful raids in the market were common and frequently condemned by the press. In January–February 1825 when the Bubble was at its peak, there were strong rumors that the Earl Grosvenor had made between £60,000 and £400,000 profit speculating in mining company shares. He denied the rumors in the House of Lords on 2 and 3 March, insisting that he had not participated in any speculation but adding that he favored bona fide companies. Furthermore, "persons who carried their surplus capital and surplus labor to other climes," he said, "were likely to do much good."[58]

Most of the banks involved in the Mexican companies survived the market crash, but there were some exceptions. The most notorious at the time was the Quaker-owned Frys and Chapman. The three principal partners were Joseph Fry, William Fry, and Jonathan Chapman. As we have indicated earlier, the Frys in particular were closely and well connected in the Quaker community being related to the Gurneys and the Buxton families. It was at least in part through such familial links that the bank acted for several of the new companies including in the Mexican group the short-lived Guanajuato Company and the better established Real del Monte Company. The bank collapsed in 1828 but the winding up and realization of assets took several years. Among those in attendance at the creditors' meetings were directors of the Real del Monte Company, the Peruvian Company, and the Brazilian

Company. At one of the meetings, creditors complained that the bank had speculated in the market but while the charge was accepted, it was pointed out that it had made a profit of £26,000 including £17,500 in Brazilian mining chares and £6,000 in Bolaños Company shares. How much money was lost by the Mexican companies who were among the bank's creditors is not known. The personal estates of the partners were sold and produced some dividends and in 1830 the life insurance policies of Joseph, aged 53, and William Fry, aged 61, both for £15,000, were sold, fetching, respectively, £4,060 and £5,160. Joseph Fry was disowned by the Quaker community for several years, only to be readmitted in 1836.[59]

Another banker who seems to have been involved in share speculation was James Heygate Jr. He was a partner in the Pares Leicestershire Bank and among other appointments, a director of the United Mexican Mining Company. In 1830, there was a major scandal when it was discovered that he had been embezzling funds from the bank. In the papers relating to this affair, now in Derbyshire Record Office, there are several financial statements of his assets and liabilities drawn up by accountants.[60] By no means easy to decipher or interpret, they nevertheless show that he was deeply involved in the United Mexican Company shares. One account indicates that he had 7,500 shares in his name; another, 400 shares at £14 each or £5,600 in total; and another, 3,800 shares. The exact significance of these details is not clear but they probably confirm that Heygate had been speculating in the share market, perhaps in the hope of returning the embezzled funds. He was also, it may be added, a heavy investor in Mexican bonds. Another Mexican company director who achieved similar notoriety at the time was Rowland Stephenson. He was a partner in the Lombard Street bank of Stephenson, Remington and Co., and a director or trustee of several companies including the Bolívar Mining Association, the Río de la Plata Mining Company, the United Provinces of Central America Mining Company, and the United Mexican Mining Company. After earlier unsuccessful attempts, he was elected to Parliament in 1826 but within two years, amidst accusations of fraud and embezzlement, his bank collapsed and he fled to the United States, where he remained for the rest of his life. His British property and assets were seized and sold but probably made little impression on the almost £500,000 of debts owed by his bank. Again, the exact cause of Stephenson's bankruptcy is not known but it is certainly possible that he too had speculated unsuccessfully in the market.[61]

Another two merchant banking partnerships closely involved in the Mexican companies also collapsed in 1826. The first was B. A. Goldsmidt and Co., contractor for the first Mexican government loan in 1824. It failed early in 1826. The financial causes of the failure—very few details are known—need not concern us but there was one effect relevant to the Mexican companies. According to Ward, the end of the Catorce Mining Company in 1829 was at least in part attributable to financial problems caused by the collapse of B. A. Goldsmidt and Co.[62] Again details of its involvement are lacking but what is known is that Adoplhus Goldsmidt was a director in 1828, his name appearing on an extant share certificate of that date.

The successful contractor for the second Mexican government loan, awarded in 1825, was the partnership of Barclay, Herring, Richardson and Co. Its leading partner was David Barclay, a merchant and member of a prominent brewing dynasty.[63] He was very active on the joint stock company scene, especially relating to Latin America, and among his appointments, he was a director of the Castello and Espiritu Santo Brazil Mining Company and the Chilian Mining Association. He was also a founding director and chairman of the Anglo-Mexican Mining Company. The Barclay partnership failed in 1826. For years afterward there were discussions with the Mexican government, which lost at least £300,000 in the failure, in an attempt to sort out the complex financial consequences. Among the multiple statements of assets drawn up in these negotiations, there is one from 1827 which shows that the Barclay firm held the following shares: 805 £100 shares in the Anglo-Mexican Company; 140 £100 shares in the Colombian Mining Association; 800 £40 shares in the United Mexican Company; 760 £25 shares in the Mexican Mine Association (this probably means the United Mexican Company): in sum, a total of 2,505 shares. There is no way of knowing the cost or value of the shares but it seems without doubt that the Barclay partnership had also been speculating in the market, especially with the Mexican companies.[64]

One final example provides a useful demonstration of the allegations of puffing up the market, alleged profiteering by directors, and the harm done to investors tempted by the "perfect moonshine" to be found in the company prospectuses.[65] The company concerned was the Mexican Company. Announced in February 1825 as both a trading and mining venture, it offered 10,000 shares with a down payment of £10 per share. There were to be fifteen directors each with a minimum of 30 shares. At first all seemed to go well. The shares quickly reached a premium of around £6, mines were contracted in Mexico, and large sums were spent on quicksilver and other mining necessities. Then the market bubble burst and by November 1825 the shares were trading at £5 or less. Angry or disillusioned shareholders soon began to make their voices heard. On 14 December, *The Times* published a detailed letter from "A Merchant." He analyzed the brief history of the company and its share price, alleging that the directors and auditors, company bankers and solicitors had all made money in the market by selling before the fall in the price. The directors have been, he said, "the great gainers" and the public should be aware of "such crying iniquities." Three days later, another letter was published from "A shareholder" and this urged that in view of what had happened to the share price and the rumors of privileged trading by the directors, a general meeting should be held to consider winding up the whole business and returning deposits on shares to investors. The directors apparently refused the request for a meeting and at the same time reminded shareholders that they were required to sign the Deed of Association before a specified date or else forfeit their shares and the money paid for them. The dissident shareholders, now increasing in number, announced that they would hold their own meeting at the George and Vulture Tavern on Saturday, 7 January 1826.

The unofficial meeting duly took place, attended by about two dozen shareholders and chaired by a Mr. Blaine.[66] Several speakers attacked the directors, accusing them of very profitable speculation in the shares, but others, including the company solicitor, Mr. Knight, denied the charge, insisting that the directors were honorable men, doing their best for the company. Unable to get their money back, the dissident shareholders adopted another tactic. They made their complaint to John Cam Hobhouse, Member of Parliament for Westminster. In several letters, he was told, with much accompanying detail, that the Mexican Company and its directors had defrauded many small shareholders out of their money while personally profiting by speculation in the shares.[67] Hobhouse was asked by one shareholder, Bernard Ross, to present a petition to the House of Commons that a committee be set up to investigate the conduct of the directors of joint stock companies in general and the Mexican Company in particular "and to take such steps as their Lordships may deem expedient to protect the Public against the future machinations of designing men." Hobhouse was clearly persuaded by the evidence put to him because on 21 February, he presented the Ross petition. The next day, another shareholder wrote to tell him of more recent allegedly illegal share transactions by the company chairman, John Mitchell, and his son. In the parliamentary debate which extended over several sessions, Hobhouse repeated the charges against the directors and while the joint stock companies in general were defended by some speakers, others, including Alexander Baring, repeated their view that the "absurd schemes" were the work of "some cunning attorney or some scheming land surveyor" and some, he added, had their origins in "the more courtly circles of the west end of town."[68]

The row went on for a few more weeks with both sides—directors and shareholders—alleging libel. The company presented a counter petition to parliament in its defense and Ross demanded that an M.P. who had libeled him in the debate should issue a retraction. Thereafter, the affair disappeared from the public scene and the Mexican Company proceeded with its operations. We have no way of knowing, of course, if the charges of speculation made against the directors were true. The extant Deed of Association reveals frequent share purchases by some directors, but as there is no indication of how much they paid or when they sold, it is impossible to calculate their profit, or loss. Since the same directors remained in charge of the company for many more years and were voted back into office repeatedly by other shareholders, it may be that the confrontation of December 1825 to March 1826 arose from the disappointment of shareholders who had failed to sell before the share price collapse.

NOTES

1. Quotation in the *Morning Chronicle*, March 1, 1826, said to be from Dryden.
2. Cobbett's *Weekly Register*, February 12 and September 17, 1825.
3. *Annual Register* (1824), 3.
4. For the investors who chose to put their money into Mexican bonds in the 1820s, see M. P. Costeloe, *Bonds and Bondholders: British Investors and Mexico's Foreign Debt, 1824–1888* (Westport, CT: Praeger, 2003), 263–99.
5. Printed by the Philanthropic Society, London.
6. Bancroft Library, University of California, BANC.Mss 67/63.
7. There are 228 entries, but some of the signatures appear more than once.
8. See Savernake Estate, Cat. Ref. 9/33/21-25.
9. Biddulph Diaries, Hereford Record Office, G2/IV/J/54.
10. Barclays Group Archives, Ref. 230/31. Copies from the share register were kindly supplied to me by the archivist, Jessie Campbell. Very difficult to decipher, they seem to show that members of the Biddulph family retained several hundred shares in the United Mexican Company until the early 1860s.
11. See the records of the Rotherwas estate and Bodenham Family, Herefordshire Record Office, Ref. AD2/IV/2.
12. See P. M. Pugh, *Calendar of the Papers of Sir Thomas Fowell Buxton, 1786–1845*, in *List and Index Society Special Series*, vol. 13, XY/N-1 (London: Swift Printers, 1980).
13. There is no evidence that Wilberforce bought shares in the joint-stock companies, but his son, William Jr., was a director of the Chilean and Peruvian Mining Company and of the Haitian Mining Company.
14. C. Buxton, ed., *Memoirs of Sir Thomas Fowell Buxton* (London: J. Murray, 1849), 244. It is also noticeable that there is no mention of Buxton's involvement in the Mexican companies in the recent *DNB* entry.
15. Swainson's correspondence is in various repositories. Some is in the Linnean Society archive, and the Museum of Zoology at Cambridge has an excellent collection of his personal papers, letters, etc. relating to his investments. It includes the very rare copies of several of the Bolaños Company reports as well as other printed papers issued by the mining companies. There is also a published collection of his letters compiled by one of his descendants. G. M. Swainson, ed., *William Swainson, FRS. FLS. Naturalist and Artist: Family Letters and Diaries, 1809–1855* (private publication, n.p. 1992). Finally, Swainson's present-day descendants still live in New Zealand and kindly allowed me to see copies of unpublished correspondence.
16. M. Swainson to John Parkes, December 18, 1827. Swainson, *William Swainson, FRS. FLS.*, 17.
17. T. S. Trail, Liverpool, to W. Swainson, May 8, 1824, Swainson Correspondence, Linnean Society, vol. 5, T-Z.
18. See J. Phillips, Real del Monte office to W. Swainson, July 25, 1832, in Swainson Papers, Museum of Zoology, University of Cambridge. One of Swainson's shares certificates is in the Chippenham collection.
19. W. Swainson, *Taxidermy: With the Biography of Zoologists* (London: Longman, 1840), 350.
20. This correspondence between Swainson and Roget is in the Swainson Correspondence, Linnean Society, vol. 4, R-S.

21. The Strutt papers are in Derbyshire Record Office, Ref. D3772/E32/121-122.

22. This letter is number LA25-002 in the Talbot Correspondence, Lacock Abbey Collection, Fox Talbot Museum, Lacock.

23. Talbot also invested in Mexican government bonds: see the letter from his broker dated October 2, 1826, Talbot Correspondence, no. 01485, and also 05846. The letters relating to his mining company shares are numbers 01335, 01369, 01373, 01382, 01791, 01801, 01807, 01809, 01811, 00129, 00278.

24. Distinctions between merchants and bankers at this time are, of course, only vague as many merchant houses carried out some banking functions and bankers were often involved in mercantile activities. See S. Chapman, *The Rise of Merchant Banking* (London: George Allen and Unwin, 1984).

25. F. G. Dawson, *The First Latin American Debt Crisis: The City of London and the 1822–1825 Loan Bubble* (New Haven, CT: Yale University Press, 1990), 141.

26. Information on Ackermann kindly supplied by John Ford.

27. For more information on McGarel and in particular his connection with the Hogg family (better known as Lord Hailsham), see Costeloe, *Bonds and Bondholders*, passim.

28. For details of Kendal investors, see ibid., 283–87.

29. Norfolk Record Office, Ref. MC 403/34/732.

30. *Trumans: The Brewers: The Story of Truman, Hanbury, Buxton, and Co. Ltd* (private publication, no author given, Newman, 1966), 22.

31. The manuscript of Taylor's unpublished letter is in the Swainson Papers, Cambridge Museum of Zoology.

32. An additional 500 shares of £400 nominal value each were issued in December 1825. They had to be paid for in £50 installments, and applications to buy from existing shareholders had priority. Michael Bland, for example, bought three shares in March 1826, numbers 583–85, on which £100 had already been paid.

33. J. Vetch, Mineral del Monte, Mexico to J. Taylor, January 24, 1826, in Swainson Papers.

34. For further information on these notables, see the *DNB*.

35. *Report Made to the Annual General Court of the Company of Adventurers in the Mines of Real del Monte, Held on the 28th of February, 1825* (London: no printer, 1825).

36. For details on William Forster, see the *DNB*. John and Jane Forster had shares in the Real del Monte Company but it is not clear if they were related to William Forster.

37. S. Finch, *Fine Meeting There Is There: 300 Years of Bridport's Quaker History* (Bridport, UK: Bridport Quaker Meeting, 2000), 72–73.

38. House of Commons debate, March 29, 1825, Hansard (1825), 1279.

39. Carnegie also invested in Mexican bonds. His son, the Reverend John Hemery Carnegie, vicar of Cranborne, near Salisbury, inherited the bonds, and in 1861 became involved in the bondholders' campaign to persuade the Mexican government to redeem them. For details, see Costeloe, *Bonds and Bondholders*, 115. It is possible that this John le Mesurier is the army officer and governor of Aldernay—see *DNB* profile for details.

40. Cited in B. Short, *A Respectable Society, Bridport, 1593–1835* (Bradford-on-Avon, UK: Moonraker Press, 1976), 54. Colfox's original letter is in the Colfox Papers, Dorset Record Office, Ref. D/COL/C20.

41. These were Colombian bonds. Letters in Colfox Papers, D/COL/B7 and B9.

42. Colfox Papers, D/COL/C20.

43. Colfox Papers, B7.

44. Colfox Papers, C29.

45. Betty Sprake is referred to in epitaphs to Thomas Colfox at the time of his death in 1835; see Colfox Papers, D/COL/F10.

46. Colfox Papers, C29.

47. R. Hills, *A Ray of Light on Mexican "Ways and Means," on the United Mexican Mining Association and on the* Quarterly Review (London: n.p., 1827), 5–15, 36.

48. Colfox Papers, B9 and C13.

49. J. L. Comstock, *A history of the precious metals . . . a history of the Anglo-Mexican Mining Companies and Speculations Concerning the Mineral Wealth of California* (Hartford, CT: Belnap and Hamersley, 1849), 198, 201.

50. English Papers, Suffolk Record Office, HA 157/3/118.

51. W. B. Pope, ed., *The Diary of Benjamin Robert Haydon* (Cambridge, MA: Harvard University Press, 1963), 3:211–12; H. McKenzie Johnston, *Missions to Mexico: A Tale of British Diplomacy in the 1820s* (London: British Academic Press, 1992), 37.

52. J. Gunn et al., eds., *Benjamin Disraeli: Letters: 1815–1834*, 8 vols. (Toronto: University of Toronto Press, c1982–c2009), 1:26–29.

53. W. F. Monypenny, *The Life of Benjamin Disraeli, Earl of Beaconsfield* (London: John Murray, 1910), 1:55; R. Blake, *Disraeli* (London: Eyre and Spottiswoode, 1966), 24–26. The records of Disraeli's stock market dealings are in the Hughenden Papers, Bodleian Library, Oxford, Box 301/2. He also invested heavily in the Latin American bond market as well as the mining companies.

54. Report in *The Hull Packet and East Riding Times*, December 27, 1844.

55. Charles Pascoe Grenfell's private ledgers are in Buckingham Record Office (Aylesbury), Ref. D/GR/9/1. Unfortunately, the pages relating to his shares in the Anglo-Mexican and United Mexican Companies have been removed.

56. These letters and documents are in the Wavell Papers in NA, E192/5.

57. Letters from Hullett Brothers to Alamán, April 9, June 10, and October 8, 1825. AGN, Hospital de Jesús, vol. 827.

58. House of Lords report in the *Times*, March 2 and 4, 1825.

59. Information on the Frys and Chapman bank is largely from reports of creditors' meetings in the *Times*, December 18, 1828, January 30, 1829, September 23, 1829, and February 4, 1830. Biographical details are from P. H. Emden, *Quakers in Commerce: A Record of Business Achievement* (London: Sampson, Low, Marston, and Co., 1939); W. H. Bidwell, *Annals of an East Anglian Bank* (Norwich, UK: A. H. Goose, 1900).

60. Derbyshire Record Office. Pares of Leicester and Hopwell Hall. Ref. GB0026 D5336.

61. For information on Stephenson, see the *DNB* entry.

62. H. G. Ward, *Mexico in 1827*, 2 vols. (London: Henry Colburn, 1828; 2nd expanded ed., 2 vols., London: Henry Colburn, 1829), 2:107–08. More details of the Goldsmidt firm's activities are in Costeloe, *Bonds and Bondholders*, passim.

63. More details on David Barclay and his activities are in Costeloe, *Bond and Bondholders*, passim.

64. The file of documents relating to the Barclay negotiations with the Mexican government is in SRE, 40-11-9. The case continued until at least 1836.

65. Most of the documentation, including the phrase "perfect moonshine," concerning the Mexican Company controversy is taken from letters and other papers in the J. C. Hobhouse, Baron Broughton Papers, British Library, ADD. Mss. 36461.

66. There is a full report of the meeting in the *Times*, January 9, 1826.

67. Broughton Papers, fols. 458–77.

68. Hansard (1826), 644–67, 698–733. The Baring speech is on 710–11.

Miscellany

We have now seen the whole range of joint stock companies set up in London between 1824 and 1860 to exploit and develop what were assumed to be the rich natural resources of independent Mexico. Reflecting long-standing commercial envy and ambitions in that part of the world, and in the whole of the Spanish American continent, Britain experienced in the 1820s a particularly strong enthusiasm for Mexico. William Bullock's notable exhibitions of Ancient and Modern Mexico, allied to his and other books, articles in the metropolitan and provincial newspapers, pamphlets by Disraeli and others, Burford's Mexican panorama, and, above all, Alexander von Humboldt's seminal work on New Spain, all combined to create a Mexicomania among the British public. This enthusiasm for the new nation coincided with and was part of the stock market Bubble of 1825 when hundreds of new joint stock companies were announced. Among them were agricultural, industrial, mining, and colonization projects in all regions of Spanish America, but it was the legendary mineral wealth of the continent that attracted the most attention. With the encouragement of Mexican mine owners and a sure belief in the advantages conveyed by superior British technology and skills, many British merchants, financiers, and entrepreneurs quickly acquired mining rights in all parts of Mexico. To raise capital to work the mines, they formed companies, and tempted by glowing prospectuses and skillful manipulation of the market by scrupulous and unscrupulous promoters, thousands of people from all parts of Britain who had money to invest bought shares. Many investors lost some or all of their money when the market Bubble, burst but in general the companies formed specifically to invest in Mexico survived. Their share prices that had risen to clearly unsustainable levels at the height of the Bubble fell dramatically but at least for two or three years afterward, they remained at a viable level that enabled the companies to begin operations in Mexico. Men were recruited and machines acquired and all were quickly dispatched across the Atlantic. Millions of pounds were spent both in Britain and in Mexico with the expectation of at least some of the shareholders that very quick profits were sure to be made. Such hopes were soon dispelled and as we have seen, the major capital investment in once famously productive mines at Guanajuato, Zacatecas, Bolaños, Catorce, and elsewhere failed to yield the expected bonanza.

By the early 1830s, if not before, it was evident that the arrival of British capital and personnel in Mexico was not going to bring quick returns. The Catorce and Tlalpujahua companies were the first to fail with the total loss of large amounts of investors' money. The five British companies that survived continued to invest and they successfully produced large amounts of gold and silver. Statistics of the value of their production are hard to come by but together with the output of Mexican-owned mines, Mexico remained the world's largest producer of silver.[1] The problem for the British companies was not production but profitability, and notwithstanding numerous experiments and improvements in mining technology that British engineers introduced, none of the companies made good profits. In a few isolated instances, for example, the Bolaños Company, there was some financial success and dividends were paid to anxious shareholders in Britain. In general, however, the financial health of all the companies was always precarious. More and more money was spent with little or no return and it was not long before shareholder patience ran out. By 1840, very little new money was leaving Britain in the direction of the five major mining companies and after approximately twenty-five years of effort and expenditure, four of the five were dissolved.

Paradoxically perhaps, the failure of the early and very well-funded British companies to make any notable profits did not mean the end of British investment in Mexico. Similarly, Mexico's chaotic and disturbing domestic political development with its characteristic features of revolution and political anarchy does not seem to have been a deterrent. Many other events that might have been expected to frighten off investors also appear to have had little effect. For example, at least until 1836 when Spain finally recognized the independence of its former colony, there was the threat of re-conquest. There was the loss of Texas in 1836; the so-called Pastry War against the French in 1838; and, above all, the devastating war with the United States from 1846 to 1848 that brought the enforced loss of almost half of Mexico's national territory. None of these conflicts convinced the British that Mexico was perhaps not the land of opportunity that they once thought. It was against this highly discouraging background in the 1830s and 1840s that several more companies were formed in London to invest in mines or in any potentially profitable venture. There was no shortage of investors willing to buy shares in the Peñoles Gold Mining Company, the Zacatecas Mining Company, the Mexican and South American Company, or the Pachuca Mining Company, all established after 1830.

Three of these companies also failed to make profits but the Mexican and South American Company was rather more successful, paying dividends to its shareholders for a number of years. It had been formed in 1835 to invest capital in any Mexican enterprise where profits could be made but as far as we know, and very little is known about the company, its main interests and investments were in the Guadalupe y Calvo area, high in the mountains of Chihuahua. It was there that the intrepid group of British adventurers led by John Hitchcock Buchan arrived in the spring of 1836 to check on the significance of the recently discovered deposits of gold and silver. Buchan promptly formed the Guadalupe y Calvo Company, attracting the support of important figures from Mexico's mercantile and financial elite. Within a couple of years, however, he and his colleagues opted to work for the Mexican and South American Company, which had bought control of some of the mines. Buchan set up house and had four children with his Mexican partner, Florencia Yañez, and several of the many other Britons who had arrived in search of work were soon well integrated into the local community. The full account of the British presence in Guadalupe y Calvo has yet to be told, and further research will almost certainly show that there is more British blood in the population than has been previously known.

Perhaps the most unusual of all the British companies examined in this present study was the hitherto unknown El Bote Mining Company. By far the most successful of the British mining ventures, it was not a joint stock company but a privately owned business, producing substantial profits for almost a century for the Kerrison and Placci families. Its survival against all the odds in the light of Mexico's turbulent history throughout the nineteenth and early decades of the twentieth centuries is nothing less than remarkable. It is further evidence that British managers, miners, and men were able to function in Mexico even during times of extremely violent civil war, for example, in the 1850s, or major revolution as in 1910–1917. Moreover, it was able to thrive even during the almost twenty years from 1867 to 1884 when Britain had no formal diplomatic relations with Mexico.

The mining industry was not the only area of Mexico's economy to attract British attention. Several of the country's mints were under British control and the Anglo-Mexican Mint Company was one of the few British-owned enterprises that was not only able to retain its concessions from successive Mexican governments but also to make handsome profits for almost half a century. Other mints in Chihuahua and elsewhere were managed by British technicians and experts in the coinage field, and much of the equipment and machinery used came from the Boulton factory in Birmingham. Also as we have shown, there was substantial British interest and investment in areas of the economy as diverse as textiles and plumbing, or cochineal and carpentry.

One area of activity in which Britons were conspicuous, but which we have only mentioned marginally, was commerce. The initial investment of those British merchants who quickly traveled to Mexico after 1821 to set up their largely import-export businesses would have come from Britain but thereafter, the investments of capital in expanding their mercantile operations or in investing in other fields such as mining or textiles came from profits generated in Mexico. No statistical evidence of capital accumulation or profits made by British merchants is available, but most economic historians have concluded that the British mercantile community was the most successful and most powerful of the various foreign and indigenous groups involved in commerce. The example of Ewen Mackintosh, and his firm of Manning and Marshall, is almost invariably used as evidence and there is no doubt that until his bankruptcy in 1851, Mackintosh and his relatives were exceptionally successful. The firm of Barron and Forbes, based at Tepic and dominant for many years on the Pacific coast, is another much-quoted example of British commercial dominance. Many other lesser-known British merchants were to be found in all the main ports and mining regions, as well as a major concentration in Mexico City.

In assessing British mercantile influence in Mexico at this time, however, some points need to be noted. The British presence, noticeable as it may have been in places like Real del Monte and Guadalupe y Calvo, was never numerically very substantial. We do not know the exact number of Britons who went and spent years living and working in the country but there were never very many. In the 1851 British Census, 755 British subjects were recorded as living in Mexico.[2] These included 458 men, 110 women, and 187 children.[3] A year earlier, in 1850, 504 had applied for security passes in Mexico.[4] There is no doubt that such low numbers were an underestimate of the total British population—perhaps they should be at least doubled—because many British migrants escaped official notice, deliberately or otherwise. British sailors, for example, jumped ship at Campeche and either disappeared or joined the Mexican army or navy. Two English carpenters working in Mexico City, Frederick Berban and James Chadley, took money in advance from their customers and then absconded with the cash.[5] Many Britons who arrived in Mexico simply disappeared, much to the anguish of their relatives back home.[6] Even if we suppose that there were as many as 1,500 Britons in the republic, it is noticeable that in the same year of 1850 when 504 had applied for security passes, 3,047 Spaniards and 803 Frenchmen applied for their cards.[7] Mexico's estimated population at the time was in the order of 7,000,000 with probably about 200,000 in the capital. In numerical terms, it is clear that the British were a negligible minority and not even a majority among the foreign population.

On the other hand, it has been argued that while not significant in number, the British merchants in particular punched much harder than their combined weight because of at least two advantages. One was the access to capital and the established contacts that they had in Britain and in some cases in Europe. Again lack of information about individual merchants and their finances precludes firm conclusions, but there is some data that appears to refute the assertion that British capital overwhelmed any opposition in the marketplace. In 1849, Francis Falconnet was in Mexico City as a representative of Barings. He studied Mexico's economy and in particular, the principal commercial and financial enterprises in the capital. He concluded that their capital value amounted to 29,000,000 pesos or £5,980,000. About 80 percent of this corresponded to Mexican- and Spanish-owned businesses. German businesses were worth around 2,500,000 pesos and French 1,500,000 pesos. In contrast, the value of British-owned companies was between 1,200,000 pesos and 1,400,000 pesos.[8]

While the situation in earlier years when more British businesses were operating in Mexico City, may have been different, it is unlikely that the relative balance, especially against Mexican- and Spanish- owned companies, was much different. At what was probably their peak in the 1830s, there were about thirty British-owned houses in central Mexico, most of which were consignment-commission establishments acting as agents or intermediaries for firms in England.[9] In 1840, there were fifty-two signatories on a tribute to the retiring British consul, John Parkinson. Fifteen were of companies, for example, Manning and Marshall, Montgomery, Nicod and Company, and Dickson, Gordon, and Co. The rest of the names were of individuals such as Ewen Mackintosh, Francis Morphy, William Phelps, and James Geaves.[10]

Although comparatively few in number, the British merchants were certainly expert at putting their case, always loudly demanding changes in legislation that affected them, for example, customs tariffs, and always vigorously protesting against any infringement of their assumed rights. In Britain, the Mexican and South American Association, a loose grouping of firms in Liverpool, Manchester, Glasgow, Belfast, and Dublin, repeatedly submitted long memoranda to the Foreign Office, claiming damages to their businesses by Mexican actions or protesting at legislation that affected their operations. Members of this Association, for example, John Diston Powles, had close links with the import-export houses throughout Mexico, and they often acted in response to information supplied to them by their partners or agents resident in Mexico.

The second alleged advantage of British merchants is the support they received from the British government. In fact, the role of the British Foreign Office and of British diplomats in Mexico is a contentious issue among some historians. From the early 1820s when Canning announced British recognition of Mexico's independence and the Treaty of Commerce and Friendship was signed by the two countries, many British merchants appealed for help.

The archive of the British Legation in Mexico City is filled with letters, memoranda, and protests from merchants around the country, and also from the British mining and other companies. Many of these appeals for diplomatic support, usually to help gain compensation from the Mexican authorities for alleged wrongdoing such as forced loans, taxes, thefts by rebels, or personal injury, were referred back to London for decision. Not all were rejected outright by the Foreign Office. Occasionally instructions were sent to the Legation to raise a matter formally with the Mexican government. For the most part, however, the British government refused to get involved. The argument was simple. Unless the issue came within the terms of a diplomatic agreement between the two countries, it could not be made a matter of official enquiry. This policy, upheld by the Foreign Office throughout the nineteenth century, was clearly enunciated on many occasions in replies to petitioners for support. In 1827, for example, the Permanent Secretary put it in these terms:

> Mr. Canning desires me in answer to state to you that it is impossible for His Majesty's Government to enter into discussion with individuals respecting the interests of their own particular speculations, or to extend favour to any of them, beyond that protection it affords alike to all British subjects and their concerns in foreign countries.[11]

A decade later, the policy was the same. In answer to a request for help in getting payment of a debt allegedly owed by the Mexican government, the petitioner Hugh McCullock, of the firm of McCullock, Hartnell, and Co., was told that:

> The British government is not in the habit of enforcing the fulfilment of contracts entered into by private individuals with Foreign Governments.[12]

Some historians, notably Tenenbaum, and those who favor the thesis that the British were formal or informal imperialists wherever they went, have argued that this official non-interventionist policy was, in fact, mere subterfuge.[13] They point to the fact that two of the most successful merchants, at least for some years, were Ewen Mackintosh and Eustace Barron, both of whom were British Consuls, a status that they argue gave them an unfair advantage over their commercial rivals. They insist that in reality, any British diplomatic pressure, even a private word behind closed doors, amounted to at least implied approval from London, and that for a Mexican to resist or deny the request was always to invite the threat or prospect of British warships appearing off the coast of Veracruz. My research for this present work, and for earlier studies, does not support the "imperialist" thesis, however informal it may have been.[14] On the contrary, the existence in the diplomatic record of so many unresolved complaints by British miners and merchants in Mexico confirms that the Mexican authorities of whatever political color, and there were dozens of changes of administration and personnel in the years between 1821 and 1860 covered by this study, did not respond to threats of any kind, direct or implied, by the Foreign Office or its representatives. Mexican bureaucrats and politicians were experts at prevarication and deflection, demonstrating in many instances negotiating skills that the British had perhaps not expected. As a result, British demands—complaints is too mild a word for many of the cases—dragged on for decades, being pursued through the courts and corridors of power on both sides of the Atlantic. It was not until the Allied Intervention and occupation of the port of Veracruz by British, Spanish, and French forces in 1862 that a few of the long-standing British claims for compensation, for example, the United Mexican Mining Company's San Acasio claim, were given official recognition. Even then, with the British and their allies in control of the customs revenues, it is by no means clear that claims were paid. Finally, it should be noted that the British government agreed to join the Allied Intervention solely on the grounds that Mexico had broken the terms of diplomatic treaties between the two countries. The private affairs and problems of British merchants and miners were not the concern of the Foreign Office.

Were British investments in Mexico a success or failure? From the viewpoint of all the companies and their shareholders, the answer to this question is clearly that the long-expected rich returns did not materialize. In short, dividends were few and far between and most investors lost their money. On the other hand, the millions of British pounds that were poured into the Mexican mines all over the country brought benefits to what had been a derelict, if not moribund industry. In areas like Guanajuato, Guadalupe y Calvo, Real del Monte, Bolaños, and Zacatecas, the mining companies were major employers, bringing a much-needed stimulus to the local economies. Improvements in mining technology were also introduced and although the British tended not to promote Mexicans to managerial positions or provide technical training, the introduction of advanced machines and innovative methods of ore extraction and refining were of obvious benefit to the industry. As far as British merchants were concerned, some made money but they were few and far between. In general, merchants remained in the country for only a few years, ending their partnerships and returning to Britain. Some, however, stayed in Mexico, married Mexican women, and established businesses that continued for several generations. As Dr. Bernecker has argued, to suggest that the British and other foreigners went to Mexico solely to exploit its natural resources and remove its wealth is manifestly incorrect. There is no evidence of repatriation of mercantile capital to Europe or elsewhere. The money that the merchants made stayed in the country to be reinvested in commerce, industry, and agriculture: "the capital accumulated by foreigners, through numerous channels, found its way into the first industrial enterprises."[15]

There were always, of course, exceptions. A few of the British migrants who began their careers in the employ of the mining companies stayed in Mexico for thirty years or more before they retired back to Britain or elsewhere in the world. Ironically, while the shareholders lost their money, these long-serving mine managers and entrepreneurs did very well financially. John Hitchcock Buchan, for example, arrived in Mexico in 1825 and finally returned to Britain in 1856. He died in 1871, leaving an estate valued at £60,000. His widow continued to enjoy handsome dividends from the Mexican-owned Real del Monte Company until her death at the fashionable resort of Biarritz in 1898. The Auld family, close friends of the Buchans, also did well. Robert Ogilvie Auld died in 1846 and we have no details of his estate but his brother, Thomas Reid Auld, who worked at Real del Monte before returning to Britain left £60,000 on his death in 1872. James Nelson Shoolbred, son-in-law of William Bullock, spent his career in Mexico working for the United Mexican Company. On his death in 1851, his estate was valued at £18,000 compared to his famous father-in-law's £650. Roger Allday Kerrison also spent over thirty years working in Mexico and he returned to his native Norfolk a very wealthy man. He left an estate worth £48,000, notwithstanding the heavy personal cost he had incurred with the bankruptcy of the Bank of Norfolk. Samuel Bradford Birkbeck is another good example. He also spent all his adult life in Mexico, retiring to Australia in 1860 with his Mexican wife and nine children. His estate at £8,000 was more modest than some but still considerable. Among the merchants who returned to Britain, some also did very well, including Edward Penny who left £30,000 and Daniel Price, £14,000. Probably the most successful in monetary terms was Alexander Forbes, the senior partner of Eustace Barron in the Pacific coast business. He retired to Scotland in 1858 and on his demise a couple of years later, his estate was valued at £50,000. Finally, there is the mysterious case of the Irish priest, Eugene McNamara. He arrived in Mexico in 1844 and was soon involved in a major colonization project for thousands of Irish immigrants to be settled in California. He was also connected to the mercury mine at New Almaden and may have held shares in it. At any rate, when he died in Paris in 1852, his estate was said to be £48,000, acquired through "mining speculations in Mexico."[16] Almost all of these men, and other examples could be given, were millionaires or multimillionaires in our twenty-first-century terms.

Such very wealthy individuals were the exception rather than the rule but they serve to illustrate that not all the British who went to Mexico left empty-handed. Although we have some idea of the number of British residents, there are no statistics of the overall total of those who visited the country. What is certain, however, is that the British never lost their fascination with its people, its culture, and its history. British visitors continued to go to all parts of Mexico, some as travelers and tourists, and others on cultural and scientific expeditions. Archaeologists examined the Mayan civilizations in the south and botanists and plant collectors made regular trips in search of new specimens.[17] There were new travelogues, plays, and poetry about the ever-popular Montezuma and the Aztecs. In 1838, for example, *Montezuma: a tragedy in five acts [and in verse]* by Dilnote Sladden was advertized. *Montezuma, a ballad of Mexico*, appeared a few years later. In 1836, *The Rambler in Mexico* by Charles Latrobe was in the bookshops.[18] He had visited Real del Monte where he observed that "an English colony has sprung up and it was moving for me to see the flaxen hair and blue eyes, and hear the prattle of many English children, gambolling in close vicinity to the swarthy offspring of the mixed races of the country." Then, in 1843 two new books were published. The first was W. H. Prescott's masterly study, *History of the Conquest of Mexico*. It was widely praised by the critics and the first London edition of 4,000 copies sold out within days. It was followed soon afterward by *Life in Mexico*, by Frances Calderón de la Barca, the Scottish-born wife of the first Spanish ambassador to Mexico. She and her husband had spent two years from late 1839 to January 1842 in Mexico, and as senior figures in diplomatic and social circles, they had met and spent much of their time with the most prominent families, including politicians, army officers, landowners, and businessmen. They also had the opportunity to travel about the country and one of the places they visited was Real del Monte. Their host was William Rule, manager of the British Real del Monte Company. After "a delicious breakfast on a snow-white table cloth, with smoking tea urn, hot rolls, fresh eggs, coffee, tea, and toast, all very much *a l'Anglaise*, the guests were taken to visit the mines. Like Latrobe before her, Frances remarked, "It was curious to see English children, clean and pretty, with their white hair and rosy cheeks, and neat straw bonnets, mingled with the little copper-coloured Indians."[19]

This image of calm, contented, British families living alongside their Mexican hosts and diligently earning their living in Real del Monte was not believed by everybody in Britain. Just two years after the publication of *Life in Mexico*, a provincial paper in Southampton, the *Hampshire Advertiser and Salisbury Guardian*, included an article entitled "Life in Mexico." Evidently with the intention of satirizing Sra. Calderón's book, it told a story, partly based on the recent murder of the Swiss Consul in Mexico City, of assassination, corruption in high places, theft, bribery, and in general, the highly dangerous nature of Mexican society. "Mexico," it affirmed, "is at the present day the only country infested with organized banditti."

How the British adapted to life in Mexico, with all its attractions and its dangers, was neatly depicted by a later visitor, William Bullock Hall, who traveled around the country in the early 1860s. Referring to English ladies he had met, he noted that they went to Church on Sundays but after the service "these ladies, such is life in Mexico, were armed with revolvers, and spent their Sunday morning in practising at a mark." Meanwhile, their husbands "devote Sunday to the healthy recreation of cricket." On Christmas Day, he added, English custom was also strictly observed with "roast beef, plum pudding, and mince pies."[20]

In sum, while the first wave of British investments in Mexico had not produced the financial returns that were expected, interest in the country's history, culture, and economy remained strong. News of the nation's domestic conflicts and wars were widely reported in the British press and death notices of British residents from violence or disease were often to be found in the London and provincial newspapers. But events, however negative, failed to diminish the British belief that Mexico was destined sooner or later to be one of the world's leading nations. *The Penny Satirist* (24 December 1842) told its readers that "Mexico is one of those countries in which nature has emptied her cornucopia and has imposed on the inhabitants little else to do than to enjoy her gifts." In a House of Commons debate in 1846, Disraeli spoke of Mexico as "a country, without any exception, of the richest resources in the world." A year later, *The Times* (9 February 1847) still referred to Mexican mines as "those seemingly exhaustless deposits of treasure." Disregarding the experience of their forbears, another generation of Britons, starting in the 1860s, was to continue the search for the elusive bonanza they were convinced was waiting to be found in Mexico's celebrated mountains of silver and gold.

NOTES

1. Cross calculated that for the fifty years from 1800 to 1850 Mexico accounted for 57 percent of all the silver mined in the world. H. E. Cross, "The Mining Economy of Zacatecas" (unpublished PhD diss., University of California, 1976), 4.
2. Number given in the *Newcastle Courant*, June 24, 1853.
3. P. Doyle to Lord Palmerston, May 5, 1851, NA, F.O. 50/244, fols. 3–4.
4. J. N. Almonte, *Guía de forasteros y repertorio de conocimientos útiles* (Mexico: Imprenta de I. Cumplido, 1852), 486.
5. *Correo de la Federación* (Mexico City), February 19, 1827.
6. See, for one of many examples, the letter to the Foreign Office from Miss Anne Williams pleading for help in finding her brother: NA, F.O. 50/242, fol. 31.
7. Almonte, *Guía*, 476.
8. D. C. M. Platt, "Finanzas británicas en México, 1821–1867," *Historia Mexicana* 32, no. 2 (1982): 243.
9. H. J. Heath, "British Merchant Houses in Mexico, 1821–1860: Conforming Business Practices and Ethics," *Hispanic American Historical Review* 73, no. 2 (1993): 268.
10. *Morning Chronicle*, February 12, 1840.
11. NA, F.O. 50/40, fol. 180.
12. Ibid., 50/121B, fol. 68.

13. B. Tenenbaum, "Merchants, Money, and Mischief. The British in Mexico, 1821–1862," *The Americas* 35 (1979): 317–39.

14. There are now many works that examine or advocate or reject the "imperialist thesis" and "dependency" interpretation of British policy and practice. The work by D. C. M. Platt, *Latin America and British Trade, 1806–1914* (London: Adam and Charles Black, 1972) remains one of the best. More recent controversial and polemical interpretations include R. D. Aguirre, *Informal Empire: Mexico and Central America in Victorian Culture* (Minneapolis: University of Minnesota Press, 2004). Another interesting recent collection of essays on South America is M. Brown, ed., *Informal Empire in Latin America: Culture, Commerce, and Capital* (Oxford, UK: Blackwell, 2008).

15. W. L. Bernecker, *De agiotistas y empresarios. En torno a la temprana industrialización mexicana (siglo XIX)* (Mexico: Universidad Iberoamericana, 1992), 190, 252.

16. McNamara's demise is reported in the *Standard*, April 19, 1852 and the *Times-Picayune* (New Orleans), November 14, 1852. Both reports were taken from the *Limerick Reporter*. For MacNamara's career, see J. Fox, *MacNamara's Irish Colony and the United States' Taking of California in 1846* (London: McFarland, 2000).

17. There are many examples of botanists collecting in Mexico in R. Desmond, ed., *Dictionary of British and Irish Botanists and Horticulturalists* (London: Taylor and Francis, 1994).

18. C. J. LaTrobe, *The Rambler in Mexico* (London: R. B. Seeley, 1836).

19. F. Calderón de la Barca, *Life in Mexico* (London: Dent, 1970), 172.

20. W. H. Bullock Hall, *Across Mexico in 1864–1865* (London: Macmillan, 1866), 116, 349.

Bibliography

MANUSCRIPT AND PRIMARY SOURCES

Full details of the archival references for the manuscript sources are given in the notes. Copies of manuscripts relating to the El Bote Mining Company were kindly supplied by Roger Kerrison. I am grateful to Melissa Boyd for copies from the Archivo de Notarías and to Linda Arnold for copies from the Archivo General de la Nación, both archives in Mexico City. The Church of All Saints Family History website provided access to many Mexican and other parish records. Copies of share certificates were generously given to me by Brian Mills. Nicholas Webb located and sent me copies of documents in Barclays Archives and Neil Boness, Margaret Ashby, and Sara Hilder found and copied many letters in the Birkbeck Papers housed in the Rare Books and Special Collections department of the Library of the University of Sydney.

NEWSPAPERS AND PERIODICALS

I have quoted from a very large number of London and provincial newspapers and periodicals but as the detailed references are given in the notes they are omitted here. Almost all are now available in the British Library's online collection of nineteenth-century newspapers and periodicals. For the U.S. newspapers, I used the online collection of *genealogybank.com.* The Mexican newspapers referred to are from my own collection of microfilm and paper copies.

BOOKS AND ARTICLES

I have divided the following list into two sections. The first, Section A, contains all those printed works cited in the text together with a small selection of other books and articles that have some relevant information. I have also included a limited selection of what is a very large number of contemporary pamphlets and other publications relating to the 1825 Bubble and to British investments in Mexico. Contemporary reviews and journals, for example, *The Edinburgh Review* and *The Gentleman's Magazine* also have many articles about Mexico but I have only mentioned those used directly in the text.

Section B contains the printed reports and other documents of the various joint stock companies. I have not included the prospectuses as these are readily available in the books by Henry English. There are also useful collections of prospectuses in the British Library and in the London Guildhall Library. Summaries of many of the later company reports are published in *The Mining Journal* and *The Quarterly Mining Review*.

A.

Abel, C., and C. M. Lewis. *Latin America: Economic Imperialism and the State: The Political Economy of the External Connection from Independence to the Present*. London: Athlone, 1985.

Adams/Rawson, W. *The Actual State of the Mexican Mines, and the Expectations of the Shareholders of the Anglo-Mexican Mine Association, Being the Substance of a Letter to the Directors of that Company*. London: Sherwood, Jones, and Co., 1825.

———. *The Present Operations and Future Prospects of the Mexican Mine Associations Analysed by the Evidence of Official Documents English and Mexican, and the National Advantages Expected From Joint-Stock Companies Considered in a Letter to the Right Hon. G. Canning*. London: J. Hatchard and Son, 1825.

Aguirre, R. D. *Informal Empire: Mexico and Central America in Victorian Culture*. Minneapolis: University of Minnesota Press, 2004.

Alamán, L. Obras. *Documentos diversos*. 3 vols. Mexico: Editorial Jus., 1942–1947.

Alanís Enciso, F. "Los extranjeros en Mexico: la inmigración y el gobierno. Tolerancia o intolerancia religiosa, 1821–1830?" *Historia Mexicana* 45, no. 3 (1996): 539–66.

Alberich, J. *Bibliografía Anglo-Hispánica, 1801–1850*. Oxford: The Dolphin Book Club, 1978.

———."English Attitudes Towards the Hispanic World in the Time of Bello as Reflected by the Edinburgh and Quarterly Reviews," 67–81 in Andrés Bello, *The London Years*, edited by J. Lynch. London: Richmond, 1982.

Alison, R. E. *To the Shareholders of the Mexican and South American Company. Reply of Mr. R. E. Alison*. London: Letts, Son, and Co., 1856.

Almada, F. R. *Guadalupe y Calvo*. Chihuahua, Mexico: [s.n.], 1940.

Almonte, J. N. *Guía de forasteros y repertorio de conocimientos útiles*. Mexico: Imprenta de I. Cumplido, 1852.

Altick, R. D. *The English Common Reader: A Social History of the Mass Reading Public, 1800–1900*. Chicago: University of Chicago Press, 1963.

An Inquiry into the Plans, Progress, and Policy of the American Mining Companies. London: John Murray, 1825.

Arellano Ramírez, L. J. *Los aventureros de minas en Bolaños: una empresa británica, 1826–1844*. Guadalajara, Mexico: Universidad de Guadalajara, 1982.
Auld R. O., and J. H. Buchan. *Notice of the Silver Mines of Fresnillo*. London: Cochrane and M'Crane, 1834.
Bagehot, W. *Lombard Street: A Description of the Money Market*. London: Henry S. King and Co., 1873.
Bazant, J. *Alienation of Church Wealth in Mexico*. Cambridge: Cambridge University Press, 1971.
Beaufoy, M. *Mexican Illustrations, Founded upon Facts with Observations upon the Government and Resources of the Republic of Mexico, as They Appeared during Part of the Years 1825, 1826, and 1827*. London: Carpenter and Son, 1828.
Bernecker, W. L. *De agiotistas y empresarios. En torno a la temprana industrialización mexicana (siglo XIX)*. Mexico: Universidad Iberoamericana, 1992.
Berninger, D. G. *La inmigración en México (1821–1857)*. Mexico: SepSetentas, 1974.
Bernstein, M., ed. *Foreign Investment in Latin America, Cases, and Attitudes*. New York: A. Knopf, 1966.
Bidwell, W. H. *Annals of an East Anglian Bank*. Norwich, UK: A. H. Goose, 1900.
Blake, R. *Disraeli*. London: Eyre and Spottiswoode, 1966.
Bonnycastle, R. H. *Spanish America*. 2 vols. London: Longman and Co., 1818.
Bottcher, N. "Casas de comercio británicas y sus intereses en América Latina, 1760–1860: estado y problemas de la investigación actual." *Ibero-Amerikanisches Archiv* 22 (1996): 191–241.
Brown, M., ed. *Informal Empire in Latin America: Culture, Commerce, and Capital*. Oxford, UK: Blackwell, 2008.
Bullock, W. *Catalogue of the Exhibition, Called Modern Mexico*. London, 1824.
———. *A Description of the Unique Exhibition Called Ancient Mexico*. London, 1824.
———. *A Descriptive Catalogue of the Exhibition Entitled Ancient and Modern Mexico*. London: printed for the Proprietors, 1825.
———. *Six Months' Residence and Travels in Mexico*. London: John Murray, 1824. Reprint, Port Washington, NY: Kennikat Press, 1971.
Bullock Hall, W. H. *Across Mexico in 1864–1865*. London: Macmillan, 1866.
Burford, R. *Description of a View of the City of Mexico and Surrounding Country*. London: J. C. Adlard, 1826.
Burkart, J. *Aufenthalt und Reisen in Mexico in den Jahren 1825 bis 1834*. 2 vols. Stuttgart, Germany: E. Schweizerbart'sche Verlagshandlung, 1826.
———. "Aufenthalt und Reisen in Mexico (Residence and Journeys in Mexico between the Years 1825 and 1834, etc.). Stuttgart, 1836. Communicated by Major Chartres." *Journal of the Royal Geographical Society* 10 (1841): 544–51.
Burt, R. *John Taylor: Mining Entrepreneur and Engineer, 1779–1863*. Buxton, UK: Moorland, 1977.
Buxton, C., ed. *Memoirs of Sir Thomas Fowell Buxton*. London: J. Murray, 1849.
Calderón de la Barca, F. *Life in Mexico*. London: Dent, 1970.
Campbell, J. *A Concise History of the Spanish America*. London: John Stagg, 1741.
Cardoso, C. F. S., ed. *Formación y desarrollo de la burguesía en México. Siglo XIX*. Mexico: Siglo Veintiuno Editores, 1978.
———. *México en el siglo XIX (1821–1910). Historia económica y de la estructura social*. Mexico: Editorial Nueva Imagen, 1981.
Chapman, S. *The Rise of Merchant Banking*. London: George Allen and Unwin, 1984.
Christelow, A. "Great Britain and the Trades from Cádiz and Lisbon to Spanish America, and Brazil, 1759–83." *Hispanic American Historical Review* 27 (1947): 2–29.
Cockburn, H. *Memorials of His Time*. Edinburgh: Adam and Charles Black, 1856.
Cody, W. F. "British Interest in the independence of Mexico, 1808–1827." PhD diss., London, 1954.
Colchester, C., ed. *The Diary and Correspondence of Charles Abbot, Lord Colchester, Speaker of the House of Commons*. 3 vols. London: John Murray, 1861.

Comstock, J. L. *A History of the Precious Metals and a History of the Anglo-Mexican Mining Companies and Speculations Concerning the Mineral Wealth of California*. Hartford, UK: Belknap and Hamersley, 1849.

Conder, J. *The Modern Traveller: A Popular Description, Geographical, Historical, and Topographical of the Various Countries of the Globe. Mexico and Guatemala*. 2 vols. London: James Duncan, 1825.

Convocatoria, que el supremo gobierno del estado de Sinaloa hace para empresarios del establecimiento de Casa de Moneda. Culiacán, 1835.

Costeloe, M. P. *Bonds and Bondholders: British Investors and Mexico's Foreign Debt, 1824–1888*. Westport, CT: Praeger, 2003.

———. "The British and an Early Pronunciamiento." In *Forceful Negotiations: The Origins of the Pronunciamiento in Nineteenth-Century Mexico*, edited by W. Fowler, 125–42. Lincoln: University of Nebraska Press, 2010.

———. *William Bullock: Connoisseur and Virtuoso of the Egyptian Hall—Piccadilly to Mexico (1773–1849)*. Bristol, UK: HiPLAM, 2008.

Coues, E., ed. *The Expeditions of Zebulon Montgomery Pike to Headwaters of the Mississippi River, through Louisiana Territory, and in New Spain, during the Years 1805-6-7*. 3 vols. London: Simpkin, Marshall, and Co., 1895.

Crick, W. F., and J. E. Wadsworth. *A Hundred Years of Joint Stock Banking*. London: Hodder and Stoughton, 1936.

Cross, H. E. "The Mining Economy of Zacatecas." Unpublished PhD diss., University of California, 1976.

Dawson, F. G. *The First Latin American Debt Crisis: The City of London and the 1822–1825 Loan Bubble*. New Haven, CT: Yale University Press, 1990.

Description of a View of the City of Mexico and Surrounding Country, Now Exhibiting in the Panorama, Leicester Square. Painted by the Proprietors, J. and R. Burford from Drawings Taken in the Summer of 1823 Brought to This Country by Mr. W. Bullock. London: J. and C. Adlard, 1826.

Desmond, R., ed. *Dictionary of British and Irish Botanists and Horticulturalists*. London: Taylor and Francis, 1994.

"Despatches Relating to British Claims." *Parliamentary Papers* 64 (1862): 85–98.

de Spix, Jean. 1926. *Serpentum Brasiliensium Species Novae*. London: Peter Davies.

Dictionary of National Biography. Oxford: Oxford University Press, 2004.

Disraeli, B. *An Inquiry into the Plans, Progress, and Policy of the American Mining Companies*. London: John Murray, 1825.

———. *Lawyers and Legislators: Or, Notes on the American Mining Companies*. London: John Murray, 1825.

———. *The Present State of Mexico: As Detailed in a Report Presented to the General Congress by the Secretary of State for the Home Department and Foreign Affairs at the Opening of the Session in 1825. With Notes and a Memoir by Don Lucas Alamán*. London: John Murray, 1825.

———. *Vivian Grey*. London: Peter Davies, 1926.

Dorset, St. John. *Montezuma: A Tragedy in Five Acts, and Other Poems*. London: Rodwell and Martin, 1822.

Doty, R. *The Soho Mint and the Industrialization of Money*. London: British Numismatic Society, 1998.

Emden, P. H. *Quakers in Commerce: A Record of Business Achievement*. London: Sampson, Low, Marston, and Co. 1939.

English, H. *A Compendium of Useful Information Relating to the Companies Formed for Working British Mines*. London: Boosey and Sons, 1826.

———. *A Complete View of the Joint Stock Companies Formed during 1824 and 1825*. London: Boosey and Sons, 1827.

———. *A General Guide to the Companies Formed for Working Foreign Mines*. London: Boosey and Sons, 1825.

Esparza, M. *Padrón general de los habitantes de la ciudad de Oaxaca, 1842*. Oaxaca, Mexico: Instituto Nacional de Antropología e Historia, 1981.

Fenn, C. *A Compendium of the English and Foreign Funds, and the Principal Joint Stock Companies*. London: Sherwood, Gilbert, and Piper, 1837.

Fenn, M. J. "British Investment in South America and the Financial Crisis of 1825–1826." Master's thesis, University of Durham, 1969.

Fenochio, A., and W. Vafeet. *Padrón general de los habitantes de la Ciudad de Oaxaca*. Oaxaca, Mexico: Instituto Nacional de Antropología e Historia, 1981.

Ferry, G. *Escenas de la vida mexicana en 1825*. Mexico: Secretaría de Educación Pública, 1945.

Finch, S. *Fine Meeting There Is There. 300 years of Bridport's Quaker History*. Bridport, UK: Bridport Quaker Meeting, 2000.

Ford, J. *Ackermann, 1783–1983: The Business of Art*. London: Ackermann, 1983.

Fox, J. *MacNamara's Irish Colony and the United States' Taking of California in 1846*. London: McFarland, 2000.

Gage, T. *A New Survey of the West Indies, or the English American, His Travel by Sea and Land*. 3rd ed. London: A. Clarke, 1677.

Gardiner, C. H. "Foreign Travelers' Accounts of Mexico, 1810–1910." *The Americas* 8 (January 1952): 321–51.

———. *Mexico, 1825–1828: The Journal and Correspondence of Edward Thornton Tayloe*. Chapel Hill: University of North Carolina Press, 1959.

Gayer, A. D., W. W. Rostow, and A. J. Schwartz. *The Growth and Fluctuation of the British Economy, 1790–1850*. Oxford: Oxford University Press, 1953.

Geldart, H. R. *Memorials of Samuel Gurney*. London: W. and F. G. Cash, 1857.

Gilliam, A. M. *Travels in Mexico during the Years 1843 and 1844*. London: Wiley and Putnam, 1846.

Gilmore, N. R. "British Mining Ventures in Early National Mexico." PhD diss., University of California, 1956.

———. "Henry George Ward, British Publicist for Mexican Mines." *Pacific Historical Review* 32 (1963): 37–38.

Goldsmith, J. *An Easy Grammar of Geography: Intended as a Companion and Introduction to the Geography on a Popular Plan for Schools and Young Persons*. London: Richard Phillips, 1805.

González Navarro, M. *Los extranjeros en México y los mexicanos en el extranjero, 1821–1970*. 2 vols. Mexico: El Colegio de Mexico, 1993.

Gregg, R. T. *Gregor MacGregor, Cazique of Poyais, 1786–1845*. London: Scripophily Library, International Bond and Share Society, 1999.

Gunn, J. G. Wiebe et al., eds. *Benjamin Disraeli Letters: 1815–1834*, 8 vols. Toronto: University of Toronto Press, c1982–c2009.

Gutiérrez, B. *Memoria sobre la creación y estado actual de las Casas de Moneda de la República*. Mexico: Tipografía de M. Murguía, 1849.

Gutiérrez López, E. O. *La inversión inglesa en la minería mexicana*. Mexico: Instituto Nacional de Antropología e Historia, 1986.

H., G. *The American Mines: Shewing [sic.] Their Importance in a National Point of View With the Progress and Present Position of the Real del Monte Company*. London: Effingham Wilson, 1834.

Haber, S. H., ed. *How Latin America Fell Behind: Essays on the Economic Histories of Brazil and Mexico*. Stanford, CA: Stanford University Press, 1997.

Hakluyt, R., ed. *The Principal Navigations, Voyages, Traffiques, and Discoveries of the English Nation*. 12 vols. Glasgow: Maclehose, 1903–1905.

Hall, B. *Extracts from a Journal Written on the Coasts of Chile, Peru, and Mexico in the Years 1820, 1821, and 1822*. 2 vols. London: Hurst and Robinson, 1824.

Hansard, T.C. The Parliamentary Debates. London: Pater-noster Row Press, 1825.

Hardy, R. W. H. *Travels in the Interior of Mexico in 1825, 1826, 1827, and 1828*. London: Henry Colburn and Richard Bentley, 1829.

Head, F. B. "Cornish Mining in America." *Quarterly Review* 36 (June 1827): 81–106.

Heath de Bohigas, H. J. "British Commercial Houses in Mexico, 1821–1867." PhD diss., London School of Economics, 1988.

———. "British Merchant Houses in Mexico, 1821–1860: Conforming Business Practices and Ethics." *Hispanic American Historical Review* 73, no. 2 (1993): 261–90.

Herrera Canales, I. *Etnía y clase. Los trabajadores ingleses de la Compañía Real del Monte y Pachuca, 1824–1906*. Mexico: Instituto Nacional de Antropología e Historia, 1981.

Hills, R. *A Ray of Light on Mexican "Ways and Means," on the United Mexican Mining Association and on the* Quarterly Review. London: n.p., 1827.

Humboldt, A. von. *Political Essay on the Kingdom of New Spain*. Translated by H. C. Woodbridge. Lexington: University of Kentucky, 1957.

Humphreys, R. A. *British Consular Reports on the Trade and Politics of Latin America, 1824–1826*. London: Royal Historical Society, 1940.

Hunt, B. C. *The Development of the Business Corporation in England, 1800–1867*. Cambridge, MA: Harvard University Press, 1936.

Informes de la junta menor permanente de la Compañía de Minas Zacatecano-Mexicana. Mexico: Impr. de I. Cumplido, 1837, 1838.

James, E. *Remarks on the Mines, Management, Ores Etc. of the District of Guanajuato, Belonging to the Anglo-Mexican Mining Association*. London: Effingham Wilson, 1827.

Jerdan, W. *Men I Have Known*. London: George Routledge and Sons, 1866.

Jiménez Codinach, G. *La comunidad inglesa en la ciudad de México*. Mexico: Instituto de la Cultura de la Ciudad de Mexico, 1999.

———. *La Gran Bretaña y la Independencia de México, 1808–1821*. Mexico: Fondo de Cultura Económica, 1991.

Kaufman, W., and H. MacPherson, eds. *Britain and the Americas: Culture, Politics, and History: A Multidisciplinary Encyclopedia*. Santa Barbara, CA: Abc-Clio, 2005.

Keen, B. *The Aztec Image in Western Thought*. New Brunswick, NJ: Rutgers University Press, 1971.

Kelly, T. *Early Public Libraries: A History of Public Libraries in Great Britain before 1850*. London: The Library Association, 1966.

Kindleberger, C. P. *Manias, Panics, and Crashes: A History of Financial Crises*. 3rd ed. Basingstoke: Macmillan, 1996.

Knight, A. "The Political Economy of Revolutionary Mexico." In *Latin America: Economic Imperialism and the State*, edited by C. Abel and C. M. Lewis. London: Athlone, 1985.

LaTrobe, C. J. *The Rambler in Mexico*. London: R. B. Seeley, 1836.

Leask, N. *Curiosity and the Aesthetics of Travel Writing, 1770–1840: "From an Antique Land."* Oxford: Oxford University Press, 2002.

Letter to John Taylor, Esq. Respecting the Conduct of the Directors of the Real del Monte Company, Relative to the Mines of Tlalpuxahua. London: Hurst, Robinson and Co. 1825.

Letters of George Borrow to the British and Foreign Bible Society. London: Hodder and Stoughton, 1911.

Liehr, R., ed. *Empresas y modernización en México desde las reformas borbónicas hasta el Porfiriato*. Frankfurt: Ibero-Americano-Vervuert, 2006.

London Post Office Directory (1819).

Ludlow, L., and C. Marichal, eds. *La Banca en México*. Mexico: Instituto Mora, 1998.

———. *Banca y Poder en México (1800–1925)*. Mexico: Editorial Grijalbo, 1985.

Lyon, G. F. *Journal of a Residence and Tour in the Republic of Mexico in the Year 1826*. 2 vols. London: John Murray, 1828.

Mackay, C. *Memoirs of Extraordinary Popular Delusions*. London: Richard Bentley, 1841.

Macmillan, D. S., and B. Plomley. "An American Surveyor in Mexico, 1827–1860." *New Mexico Historical Review* 34, no. 1 (January 1959): 1–8.

Márquez Terrazas, Z. *Pueblos Mineros de Chihuahua*. Chihuahua, Mexico: Universidad Autónoma de Chihuahua, 2007.

Marshall, O., ed. *English-Speaking Communities in Latin America*. London: Palgrave, 2000.

Martineau, H. *A History of the Thirty Years' Peace, 1816–1846*. 3 vols. London: Charles Knight, 1849–1850. Reprint, Shannon: Irish University Press, 1971.

Mateos, J. A., ed. *Historia parlementaria de los congresos mexicanos*. 25 vols. Mexico: Imprenta de J.F. Jens et al., 1877–1912.

Mayo, J. *Commerce and Contraband on Mexico's West Coast in the Era of Barron, Forbes, and Co., 1821–1859*. New York: Lang Peter, 2006.

McCulloch, J. R. *Dictionary of Commerce and Commercial Navigation*. London: Longman, Rees, Orme, and Green, 1834.

———. *A Dictionary, Practical, Theoretical, and Historical, of Commerce and Commercial Navigation*. 2 vols. London: Longman, Rees, Orme, Brown, and Green, 1840–1843.

McKenzie Johnston, H. *Missions to Mexico: A Tale of British Diplomacy in the 1820s*. London: British Academic Press, 1992.

Meyer Cosio, R. M. "Los ingleses en Mexico. La casa de Manning y Mackintosh (1824–1852)." *Historias* 16 (1987): 57–71.

———. *Los inmigrantes en el mundo de los negocios*. Mexico: Instituto Nacional de Antropología e Historia, 2003.

———. "La quiebra de la firma Manning y Mackintosh." In *Prestigio, riqueza y poder. Las élites en México, 1821–1940*, edited by G. Altamirano, 87–114. Mexico: Instituto Mora, 2000.

Michie, R. C. *Money, Mania, and Markets: Investment, Company Formation, and the Stock Exchange in Nineteenth-Century Scotland*. Edinburgh: Donald, c. 1981.

Mill, N. *The History of Mexico: Also Observations as to Working the Mexican Mines by British Capital*. London: Sherwood, Jones, and Co., 1824.

Miller, R. *Britain and Latin America in the Nineteenth and Twentieth Centuries*. London: Longman Books, 1993.

The Mining Journal, Railway, and Commercial Gazette (1835–1860).

Monroy Castillo, I. *Sueños, tentativas y posibilidades. Extranjeros en San Luis Potosí, 1821–1845*. San Luis Potosí, Mexico: El Colegio de San Luis Potosí, 2004.

Montejano y Aquinaga, R. *El Real de Minas de la Purísima Concepción de los Catorce*. San Luis Potosí, Mexico: Academia de Historia Potosina, 1975.

Montesinos y Aquinaga, R. *El Real de Minas de la Purísima Concepción de los Catorce*. San Luis Potosí, Mexico: Academia de Historia Potosina, 1975.

Monypenny, W. F. *The Life of Benjamin Disraeli, Earl of Beaconsfield*. London: John Murray, 1910.

Muhlenpfordt, E. *Ensayo de una fiel descripción de la República de México*. Mexico: Banco de Mexico, 1993.

———. *Los palacios de los zapotecas en Mitla*. Mexico: UNAM, 1984.

Neal, L. "The Financial Crisis of 1825 and the Restructuring of the British Financial System." Prepared for the 22nd Annual Economic Policy Conference at the Federal Reserve Bank of St. Louis, October 16–17, 1997, www.business.uiuc.edu/lneal?Econ238/gold.htm.

Niles, J. M. *View of South America and Mexico*. Montreal: A. G. and A. D. Collins, 1827.

Nzibo, Y. A. "Relations between Great Britain and Mexico, 1820–1870." PhD diss., Glasgow, 1979.

Olveda, J., ed. *Inversores y empresarios extranjeros en el noroccidente de México. Siglo XIX*. Jalisco, Mexico: El Colegio de Jalisco, 1996.

Pagden, A. *European Encounters with the New World from Renaissance to Romanticism*. New Haven, CT: Yale University Press, 1993.

Parra, A. *Estado y Minería en México, 1767–1910*. Mexico: Fondo de Cultura Económica, 1988.

———. *La presencia inglesa en México*. Mexico: Instituto Nacional de Historia, 1995.

Pawlyn, T. *The Falmouth Packets, 1689–1851*. Truro, UK: Truran Books, 2003.

Pedraza, J. F. "Robert Phillips. Detalles de un viaje desde Altamira a Catorce." *Archivos de Historia Potosina* 5, no. 2 (October–December 1973): 104–40.

Phillips, R. *A Million of Facts*. London: Darton and Hodge, 1832.

Pitman, R. B. *A Succinct View and Analysis of Authentic Information Extant in Original Works on the Practicability of Joining the Atlantic and Pacific Oceans, By a Ship Canal across the Isthmus of America*. London: J. M. Richardson, J. Hatchard, and Son, 1825.

Placci, G. *Statement of Facts Relative to the Negotiation of Cerro del Bote, in the State of Zacatecas, Mexico*. London: A. Mackintosh, 1838.

Plan de reorganización de la Compañía de Minas denominada Restauradora de Mineral de Catorce. Mexico: Ignacio Cumplido, 1851.

Planché, J. R. *Cortez: Or, The Conquest of Mexico*. London: John Lowndes, 1823.

Platt, D. C. M. *Business Imperialism, 1840–1930. An Inquiry Based on British Experience in Latin America*. Oxford: Oxford University Press, 1977.

———. "Finanzas británicas en México, 1821–1867." *Historia Mexicana* 32, no. 2 (1982): 226–61.

———. *Latin America and British Trade, 1806–1914*. London: Adam and Charles Black, 1972.

Plomley, N. J. B. *Several Generations*. Sydney: Wentworth, 1971.

Poinsett, J. R. *Notes on Mexico*. Reprint, New York: Frederick A. Praeger, 1969.

Polushin, M. A. "British Investment in Mexican Mining, 1824–1828." MA thesis, University of Alberta, 1990.

Pope, W. B., ed. *The Diary of Benjamin Robert Haydon*. 5 vols. Cambridge, MA: Harvard University Press, 1963.

Potash, R. *Mexican Government and Industrial Development in the Early Republic: The Banco de Avío*. Amherst: University of Massachusetts Press, 1983.

Powles, J. D. *A letter to Alexander Baring, Esq. M. P. on the Subject of Some Observations Reported to Have Been Made by Him in the House of Commons, on the 16th March, 1825, in Relation to the Foreign Mining Associations*. London: Cowie and Co. 1825.

Probert, A. "Silver Quest: Episodes of Mining in New Spain—Nine Readings." *Journal of the West* 14, no. 2 (1975): 1–168.

Pugh, P. M. *Calendar of the Papers of Sir Thomas Fowell Buxton, 1786–1845*. Vol. 13 of List and Index Society Special Series. London: Swift Printers, 1980.

The Quarterly Mining Review. 4 vols. London: T. and T. Boosey, 1830–1837.

Rabadan Figueroa, M. *Propios y extraños: la presencia de los extranjeros en la vida de la Ciudad de México, 1821–1860*. Mexico: El Colegio de Mexico, 2000.

Racine, K. "Imagining Independence. London's Spanish American Community, 1790–1829." Unpublished PhD diss., Tulane University, 1996.

Randall, R. W. *Real del Monte: A British Mining Venture in Mexico*. Austin: University of Texas Press, 1972.

Rankine, M. E. "The Mexican Mining Industry in the Nineteenth Century with Special Reference to Guanajuato." *Bulletin of Latin American Research* 11 (1992): 29–48.

———. "The Mexican Silver Mining Industry in the Nineteenth Century with Special Reference to Guanajuato." PhD diss., University of Cambridge, 1986.

Richardson, C. *Mr. John Diston Powles: Or the Antecedents as a Promoter and Director of Foreign Mining Companies or an Administrative Reformer*. London: printed for the author, 1855.

Rio, A. del, and P. F. Cabrera. *Description of the Ruins of an Ancient City Discovered Near Palenque in the Kingdom of Guatemala in Spanish America*. London: H. Bertoud, 1824.

Rippy, J. F. *British Investments in Latin America*. London: Routledge, 2000.

———. "Latin America and the British Investment Boom of the 1820s." *Journal of Modern History* 19 (June 1947): 122–29.

Robertson, W. *The History of America*. 3 vols. London: W. Strahan, T. Cahill, and J. Balfour, 1777.

Robertson, W. P. *A Visit to Mexico*. London: Simpkin, Marshall, and Co., 1853.

Robinson, W. D. *Memoirs of the Revolution of Mexico, with a Narrative of the Campaign of General Mina, Anecdotes of His Life, and Observations on the Practicability of Connecting the Pacific with the Atlantic Ocean, by Means of Navigable Canals*. London: Leckington, Hughes, and Co., 1821.

Rolt, R. *A New and Accurate History of South America*. London: printed for T. Gardner, 1756.

Romney, A. (pseud?) *Three Letters on the Speculative Schemes of the Present Times, and the Projected Banks. Addressed to His Cousin Richard, in the Country*. Edinburgh: Bell and Bradfute, 1825.

Rowley, G. D., ed., *Ornithological Miscellany*. 3 vols. London: Trubner, 1876–1878.

Salvucci, R. *Politics, Markets, and Mexico's "London Debt," 1823–1887*. New York: Cambridge University Press, 2009.

Schwartz, S. "Creating the Cult of 'Cousin Jack': Cornish Miners in Latin America, 1812–1848 and the Development of an International Mining Labour Market." www. projects.exeter.ac.uk/ cornishlatin/Creating%20the%20Cult%20of%20Cousin%20Jack.pdf.

Selincourt, E. de, ed. *The Letters of William and Dorothy Wordsworth: The Later Years*. Oxford: Oxford University Press, 1939.

Sevin, C. "A Journey to Mexico". *Journal of the Royal Geographical Society* 30 (1860): 1–53.

Short, B. *A Respectable Society, Bridport, 1593–1835*. Bradford-on-Avon, UK: Moonraker Press, 1976.

Skeat, W. D. *George Stephenson: The Engineer and His Letters*. London: Institute of Mechanical Engineers, 1973.

Smith, M. D. R. *Recollections of Two Distinguished Persons: La Marquise de Boissy and the Count of Waldeck*. Philadelphia: J. B. Lippincott and Co., 1878.

Smith, R. A. *Borderlander: The Life of James Kirker, 1793–1852*. Norman: University of Oklahoma Press, 1999.

Staples, A. *Bonanzas y Borrascas. Mineras. El Estado de México, 1821–1876*. Mexico: El Colegio Mexiquense, 1994.

Steele, C. *English Interpreters of the Iberian New World from Purchas to Stevens: A Bibliographical Study, 1603–1726*. Oxford, UK: Dolphin Book Company, 1975.

Stein, J., and B. H. Stein. *The Colonial Heritage of Latin America*. New York: Oxford University Press, 1970.

Strachey, L., and R. Fulford. *The Greville Memoirs, 1814–1860*. London: Macmillan, 1938.

Stresemann, E. *Ornithology from Aristotle to the Present*. Cambridge, MA: Harvard University Press, 1875.

Swainson, G. M., ed. *William Swainson, FRS. FLS. Naturalist and Artist: Family Letters and Diaries, 1809–1855*. Private publication, n.p., 1992.

Swainson, W. *Taxidermy: With the Biography of Zoologists*. London: Longman, 1840.

Sweet, J. M. "William Bullock's Collection and the University of Edinburgh, 1819." *Annals of Science* 26, no. 1 (1970): 23–32.

Taylor, J. *Selections from the Works of Baron de Humboldt, Relating to the Climate, Inhabitants, Productions, and Mines of Mexico*. London: Longman, Hurst, et al., 1824.

———. *Statements Respecting the Profits of Mining in England, Considered in Relation to the Prospects of Mining in Mexico; in a Letter to Thomas Fowell Buxton, M.P.* London: Longman and Co., 1825.

Tenenbaum, B. "Merchants, Money, and Mischief. The British in Mexico, 1821–1862." *The Americas* 35 (1979): 317–39.

Tenenbaum, B. A., and J. N. McElveen, "From Speculative to Substantive Boom: The British in Mexico, 1821–1911." In *English-Speaking Communities in Latin America*, edited by O. Marshall. London: Palgrave MacMillan, 2000.

Todd, A. C. *The Search for Silver: Cornish Miners in Mexico, 1824–1947*. Padstow: The Lodenek Press, 1977. Reprint, Exeter, UK: Short Run Press, 2000.

Trewman's Exeter Flying Post, April 27, 1870.

True, C. A. "British Economic Interests and Activities in Mexico, 1830–1846." PhD diss., University of California, 1933.

Trumans: The Brewers: The Story of Truman, Hanbury, Buxton and Co. Ltd. London: private publication, no author given, Newman, 1966.

Valadés, J. C. *Alamán. Estadista e Historiador*. Mexico: Universidad Nacional Autónoma de México, 1977.

Velasco Avila, C. et al. *Estado y Minería en México (1767–1910)*. Mexico: Fondo de Cultura Económica. 1988.

———. "Labour Relations in Mining: Real del Monte and Pachuca, 1824–1874." In *Miners and Mining in America*, edited by T. Geaves and W. Culver, 47–67. Manchester: Manchester University Press, 1985.

Velázquez, R. V. *Inmigrantes británicos en el distrito minero de Real del Monte y Pachuca, 1824–1947: un acercamiento a la vida cotidiana*. Mexico: Archivo Histórico y Museo de Minería, 2004.

Villegas Revueltas, S. *Deuda y Diplomacia. La relación México-Gran Bretaña, 1824–1884*. Mexico: Universidad Nacional Autónoma de México, 2005.

Vincent, J. *An Intelligent Person's Guide to History*. London: Duckworth, 1995.

Walker, D. W. *Kinship, Business, and Politics: The Martínez del Río family in Mexico, 1823–1867*. Austin: University of Texas Press, 1986.

Walpole, S. *A History of England from the Conclusion of the Great War of 1815 (to 1840)*. London: Longmans, Green, and Co., 1878–1886.

Walton, W. *The Present State of the Spanish Colonies*. 2 vols. London: Longman, Hurst, Rees, Orme, and Brown, 1810.

Ward, H. G. *Mexico in 1827*. 2 vols. London: Henry Colburn, 1828; 2nd expanded ed., 2 vols. London: Henry Colburn, 1829.

Waterston, W. A. *Cyclopaedia of Commerce, Mercantile Law, Finance, Commercial Geography, and Navigation*. London: Henry G. Bohn, 1844.

Weiss, B. *The Hell of the English: Bankruptcy and the Victorian Novel*. London: Associated University Presses, 1986.

Williford, M. *Jeremy Bentham on Spanish America: An Account of His Letters and Proposals to the New World*. Baton Rouge: Louisiana State University Press, 1980.

Willis, R. J. *Justus Ludewig von Uslar, and the First Book on Alleopathy*. Dordrecht, The Netherlands: Springer, 2004.

B.

Company Reports and Miscellaneous Publications

This is not a comprehensive list of all publications by the companies but of those that I have used. It is probable that copies of many of the yearly and half-yearly reports have not survived. *The Quarterly Mining Review* (1830–1837) has good summaries of the later meetings for which the original printed reports are not extant. *The Mining Journal* has summaries of the meetings from 1835 onwards and most of the London newspapers have some details.

Anglo-Mexican Mining Company

Report of the Directors of the Anglo-Mexican Mining Association, presented at the Special General Meeting of the Proprietors, held at the City of London Tavern, Monday, August 28, 1826. London: Henry Stokes, 1826.

Report of the Directors of the Anglo-Mexican Mining Association, presented at the Second Annual General Meeting of the Proprietors, held at the City of London Tavern, on Wednesday, the 3rd of January, 1827. London: Richard Clay, 1827.

Report of the Directors of the Anglo-Mexican Mining Association, presented at the Third Annual General Meeting of the Proprietors, held at the City of London Tavern, on Wednesday, the 3rd of January, 1828. London: R. Clay, 1828.

Report of the Directors of the Anglo-Mexican Mining Association, presented at the Fifth Annual General Meeting of the Proprietors held at the City of London Tavern, on Wednesday, July 1, 1829. London: R. Clay, 1829.

Bolaños Company

The Bancroft Library has a comprehensive collection of copies of the Bolaños Company reports of yearly and half-yearly meetings of shareholders from 1827 to 1845, including special reports prepared for the meetings.

Mexican Company

Report of the Directors of the Mexican Company presented at the Annual General Meeting of Proprietors, held at the City of London Tavern, on Thursday, the 4th of May, 1826. London: J. Plummer, 1826.

Report of the Directors of the Mexican Company, presented at the Second Annual General Meeting of the Proprietors, held at the City of London Tavern, on Thursday, 3rd of May, 1827. London: J. Plummer, 1827.

Mexican Company. At a General Meeting of Proprietors, held at the City of London Tavern (in conformity with the Deed of Association) on Thursday, the 1st of May, 1828. London: J. Plummer, 1828.

Uslar, J. L. Von. *Extracts from the report of Mr. Justus Ludwig Von Uslar (Chief Director of the mines in the service of the Mexican Company) relative to the "Negociación" of Yavesía in the state of Oaxaca. Dated San Antonio de Padua (Oaxaca), 6th January, 1828.* London: J. Plummer, 1828.

Report of the Directors of the Mexican Company, presented at the Fifth Annual General meeting of the Proprietors, held at the City of London Tavern, on Thursday, 6th May, 1830. London: J. Plummer, 1830.

Mexican and South American Company

Extracts from the advices received by the Directors, since the Seventh Annual General Meeting, held on the 8th June, 1842, printed for the information of the proprietors and presented at the Eighth Annual General Meeting held at No. 9, New Broad Street, 14th June, 1843. London: R. Clay, 1843.

Extracts from the advices received by the Directors since the Eighth Annual General Meeting, held on 14th June, 1843, printed for the information of the Proprietors, and presented at the Ninth Annual General Meeting, held at No. 5, Broad Street Buildings, 12th June 1844. London: R. Clay, 1844.

Extracts from the Advices received by the Directors, since the Ninth Annual General Meeting, held on the 12th June, 1844, printed for the information of the Proprietors, and presented at the Tenth Annual General Meeting, held at No. 5 Broad Street Buildings, 11 June, 1845. London: R. Clay, 1845.

Peñoles Gold Mining Company

Report of the Directors of the Peñoles Gold Mining Association to be presented at the fourth half-yearly general meeting of the proprietors which will be held at the office of the Association on Monday, the 6th of May 1833. London: E. B. and G. Clarke, 1833.

Report [dated May 4, 1835]. London: T. Smith, printer, 1835 (also contains report of meeting of November 3, 1834.

Real del Monte Mining Company

Report made to the Annual General Court of the Company of Adventurers in the Mines of Real del Monte, held on the 28th of February, 1825. London: n.p., 1825.

Proceedings at the several general Courts of Proprietors of the Real del Monte Mining Company, since the Annual General Court held 28th February, 1825. London: n.p., 1826.

Taylor, J. *Report made to the Annual General Court of the Company of Adventurers in the Mines of Real del Monte, 28th February, 1826.* London: n.p., 1826.

Tlalpujahua Company

Report of the Directors, 6 April 1826. London: n.p., 1826.

Burkart, J. *Report containing the result of the preliminary examination of the mines in the possession of the Real de Tlalpuxahua Mining Association.* London: W. Marchant, 1826.

United Mexican Mining Association

Report of the Court of Directors addressed to the shareholders. London: printed by the Philanthropic Society, 1825.

Report of the Special General Meeting held at the City of London Tavern, the 19th July 1826. London: Philanthropic Society, 1826.

Circular of the Court of Directors of the United Mexican Mining Association addressed to the Proprietors. Dated London, 11 November, 1826. London: printed by the Philanthropic Society, 1826.

Report of Don Lucas Alamán, addressed to the Directors; dated at Mexico, the 28th May, 1826, and a letter from Baron de Humboldt, addressed to the Directors. London: printed by the Philanthropic Society, 1826.

Report of the Court of Directors, addressed to the Proprietors at a General Meeting on the 7th March, 1827. London: printed by the Philanthropic Society, 1827

Report of the Court of Directors, dated 13th June, 1827. London: printed by the Philanthropic Society, 1827.

Report of Proceedings at a General Meeting of Proprietors, 25th July, 1827. London: printed by the Philanthropic Society, 1827.

Report of the affairs of the United Mexican Mining Association, made to the Court of Directors, by the Secretary, on his return from Mexico. Dated 31st May, 1828. London: printed by the Philanthropic Society, 1828.

Substance of information received by the Court of Directors of the United Mexican Mining Association since the return of the secretary from Mexico. Dated 14th June, 1828. London: printed by the Philanthropic Society, 1828

Report of Proceedings at a General Meeting of Proprietors, 30 July 1828. London: printed by the Philanthropic Society, 1828.

List of Proprietors of the United Mexican Mining Association, corrected to the 29th July, 1829. London: printed by the Philanthropic Society, 1829.

Report of Proceedings at a General Meeting of Proprietors, 17th March, 1830. London: Baynes and Harris, 1830.

Report of a General Meeting of Proprietors, 6th August, 1830. London: Baynes and Harris, 1830.

Report of a General Meeting of Proprietors, 26th January 1831. London: Baynes and Harris, 1831.

Report of a General Meeting of Proprietors, 7th June, 1831. Baynes and Harris, 1831.

Report, 27th July 1831. London: Baynes and Harris, 1831.

Report of a Committee of Proprietors, read at a Special General Meeting, 18th of October, 1831. London: Baynes and Harris, 1831.

List of the Proprietors of the United Mexican Mining Association. London: Baynes and Harris, 1831.

Report, 14th March, 1832. London: Baynes and Harris, 1832.

List of the Proprietors of the United Mexican Mining Association. London: R. Clay, 1833.

Report, 31st July, 1833. London: R. Clay, 1834.

List of the Proprietors of the United Mexican Mining Association. London: R. Clay, 1834.

Index

CPSIA information can be obtained at www.ICGtesting.com
262511BV00004B/1/P

9 780739 151198